Do-It-Yourself Home Energy Audits

140 Simple Solutions to Lower Energy Costs, Increase Your Home's Efficiency, and Save the Environment

David S. Findley

New York Chicago San Francisco
Lisbon London Madrid Mexico City
Milan New Delhi San Juan
Seoul Singapore Sydney Toronto

The McGraw·Hill Companies

Cataloging-in-Publication Data is on file with the Library of Congress

McGraw-Hill books are available at special quantity discounts to use as premiums and sales promotions, or for use in corporate training programs. To contact a representative, please e-mail us at bulksales@mcgraw-hill.com.

Do-It-Yourself Home Energy Audits

1 2 3 4 5 6 7 8 9 0 DOC/DOC 1 6 5 4 3 2 1 0

ISBN 978-0-07-163639-1
MHID 0-07-163639-0

The pages within this book were printed on acid-free paper containing 100% postconsumer fiber.

Sponsoring Editor
Judy Bass

Editorial Supervisor
Stephen M. Smith

Production Supervisor
Pamela A. Pelton

Acquisitions Coordinator
Michael Mulcahy

Project Manager
Patricia Wallenburg, TypeWriting

Copy Editor
James Madru

Proofreader
Paul Tyler

Indexer
Judy Davis

Art Director, Cover
Jeff Weeks

Composition
TypeWriting

This book is dedicated to the four most important people in my life:

My girlfriend, for always and all ways loving me;

My mother, for her love and unconditional support;

My sister, who is always willing to help; and

My father, who recently reduced his CO_2 emissions
to zero—you will be forever missed.

About the Author

David S. Findley is an assistant professor at Farmingdale State College, Farmingdale, New York, and the owner of Synergy New Technology, a not-for-profit green technology solution provider. He has overseen the development of a green engineering sciences curriculum.

Contents

Preface

It is my sincerest hope to be of assistance to you. My hope is to keep you warmer in the winter and cooler in the summer and to help you design and live in the home of your dreams. I also hope to save you thousands of dollars each year.

I will use the word *home* for the place where you live. It does not matter if you live in a one-room apartment or a one-hundred-room mansion—this book is for you. The changes shown to you in this book will allow you to reduce your energy usage and increase your savings by at least, on average, 30 percent.

I will show you the small changes you can make today and how to plan the renovation of your entire home. This book includes many energy- and money-saving ideas that you can implement at no cost.

I provide you with a host of resources that are available for free. You will be able to get more information about your home projects, green technology, or climate change.

This brings us to my final subject: *You can improve your life while saving the world.* While this idea may sound outlandish, it is true. For the first time in history, you can be a superhero (costume optional). How is this possible? You will need to read on.

My hope is that together we can improve your life and help you to save a significant portion of your income. The added benefit of your actions will be a better world tomorrow for the children of today. Please send me your ideas and comments at HEA@exploresynergy.org.

David S. Findley

The 140 Simple Solutions

Chapter 1

Chapter 2

Chapter 3

Chapter 4

Chapter 5

Chapter 6

Chapter 7

Chapter 8

Chapter 9

Chapter 10

CHAPTER 1

Increase Your Home's Comfort and Increase Your Income

Welcome to better living. Most people do not realize how much time they spend in their homes. We can spend as much as 90 percent of our time indoors. Why would you want to pay extra for that privilege? Most homeowners pay $2,500 per year or more for utilities. Does your home feel as comfortable as you would like? Would you like to enjoy and be content and relaxed every moment that you are in your home? I know that I would. So how can you achieve this level of comfort and relaxation? This may seem obscure, but the answer is a technical one—by making your home more efficient.

If you could increase both your comfort and your income, does that sound like a solution that you would enjoy? I will guide you through each part of your home. I will assist you in making your home more energy efficient. It does not matter if you live in a house, an apartment, or any other type of dwelling. I will assist you to identify the problems in your home, help you to eliminate the waste, and, hopefully, put a smile on your face.

I will show you the total cost of ownership of your home, that is, how much it actually costs you to live in your present home. I will use an average house for this example. If you own a home, you probably pay a mortgage. If you financed your home correctly, the cost of your home should be only about one-third of your total income. Where does the rest of your income go? You will find out below.

The Relationship Between Your Home and Your Wallet

Let's begin by demonstrating the interrelationships among the systems in your home. I will use the simple act of turning on the hot-water faucet in your kitchen as a starting place. Turn the hot-water faucet to the on position, and the water slowly becomes warm and then hot. The first expense is the water. The water is being heated by the hot-water heater as part of the furnace. A boiler is the home heater that also produces the hot water when no separate hot-water heater is present. The hot-water heater uses either electricity or electricity and gas to heat the water. The simple operation of hand washing has used three utilities (water, electricity, and gas). This is a simple need but a costly and complex solution. Does it make sense? The answer has evolved to become unequivocally, "No." We will learn why when you learn how energy is produced and used in your home.

So how did we get here? Until recently, coal and oil prices have been low. We have used vast amounts of oil and coal to power our inefficient homes and our lives. Homes built even today lack an effective and efficient amount of mass or insulation.

Despite the media frenzy about green energy, clearly the United States and the world are still completely dependent on fossil fuels. Most people do not realize that we spend most of our life indoors. Our jobs, schooling, and even dining out usually occur inside one kind of building or another besides our homes. What most people are beginning to ask is, "Is there a better way to live?" The answer is unequivocally, "Yes." What most people do not grasp is that the change to a better world is the choice of individuals. When you reduce your waste, buy green energy, take part in your government, and make new forms of energy profitable, then change will occur.

Individually, these changes can be made over time and without significant cost or effort. I will provide ideas that will cost no money to implement. I will help you to understand how you live currently and how much better you could be living. *Warmer, safer, wealthier*—are these good adjectives? What will this cost you? Deliberation, reflection, and review of how you presently live and how you can benefit from small changes in your life.

Immediate No-Cost Home Energy Solutions

I will provide some immediate cost-saving projects for you to accomplish in each chapter. So let us begin right now with your first opportunities to save money.

Analyze Your Home Energy Expenses

Most people worldwide do not reflect on how they live. Most people do not plan for life. If people do not plan and do not reflect on their actions, how can they improve their lives? Let us begin by reflecting on your life in terms of monthly expenses. Table 1-1 illustrates the average monthly expenses of a suburban homeowner. I use a suburban homeowner as an example because suburban homes are the least-efficient energy consumers. While cities use much more energy, apartments and apartment complexes house more people and use less energy.

TABLE 1-1 Individual Suburban Homeowner Expenses

Monthly Expenses	**Current**	**Adjusted**
Taxes	$600.00	$600.00
Water	$25.00	$0.00
Electric	$200.00	$0.00
Heat	$300.00	$0.00
Health insurance	$500.00	?
Medical	$30.00	?
Food	$300.00	$250.00
Car	$400.00	$100.00
Gas	$200.00	$0.00
Maintenance	$100.00	$10.00
Car insurance	$100.00	$50.00
Cable	$50.00	$50.00
Internet	$40.00	$40.00
Totals	$2,755.00	$1,100.00
Savings each month		**$1,655.00**

The table lists average costs for a typical homeowner in any suburban area. These expenses include health insurance and homeowner's insurance. These two expenses can be reduced with clean living and tax incentives and rebates.

Are such large savings possible? The answer is yes. Choosing to address the chief energy issues in your home potentially could result in more expendable income. Many of the energy-efficient solutions offered here will cost you little or no money. Before I begin with global warming, how the electrical grid works, and the exciting products called *insulation*, let's start with a few things that will cost you no money and reduce your expenses immediately.

Your income pays for all the related home expenses. Your expenses probably amount to as much as your mortgage. You cannot reduce the amount of your mortgage, but you can reduce the amount of your home expenses. The total cost of homeownership is a total of many expenses. Almost all energy-efficient home improvements, from fixes for leaky faucets to new windows, will pay for themselves in time.

Use Your Current Natural Ventilation

The use of a function that requires no energy input is always preferable. Your home does not require a redesign to use natural ventilation. Most homes allow air to flow freely. Figure 1-1 demonstrates natural airflow and cooling with no energy use. The problem is that most homes act in the same manner when containment of the home's air is actually what is required.

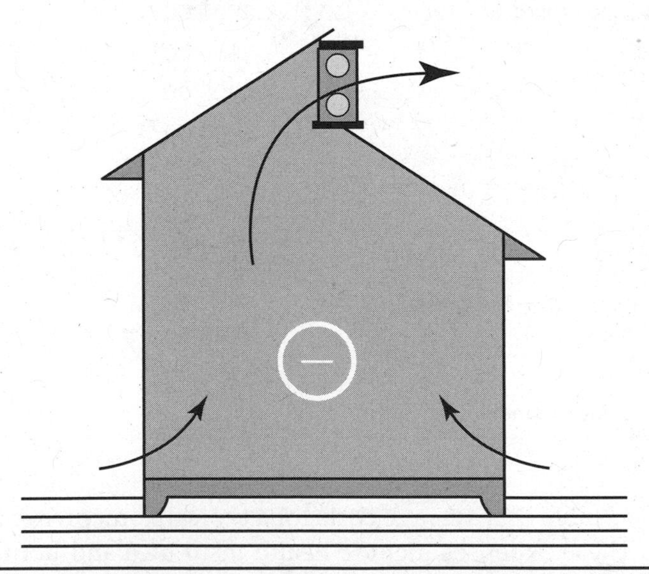

FIGURE 1-1 Natural airflow in a home. *(Louisiana Department of Natural Resources, http://dnr.louisiana.gov.)*

FIGURE 1-2 Natural ventilation. *(National Renewable Energy Laboratories, www.nrel.gov/.)*

In the Northeast, my neighbor turns on his air conditioner in March and turns it off in December. Despite the sometimes freezing weather, he uses the air conditioner as a ventilator. He often has his air conditioner and heat on simultaneously. Why? He and his wife feel that it is difficult to open and close the windows in their home. Instead, they choose—and make no mistake, this is a choice—to pay $400 per month for electricity.

Fresh, clean air is supplied to your home for free and there is no need to pump the air in when you simply can let it in. In temperate weather, opening windows will provide natural ventilation (Fgure 1-2). This will solve indoor air pollution problems and save a great deal of money. If you live in a one-room apartment with a single window, then Figure 1-3 shows how easy it is to manage your natural ventilation.

Use Your Curent Natural Lighting

Open your blinds, open your windows, and use natural light. Much like your air, light is delivered to your home for free; you just need to let it in to your home. Figure 1-4 demonstrates how natural light can be used to supplement or replace electric light. Figure 1-5 shows rooms that require no additional light all day.

Figure 1-3 Window ventilation. *(Fairfax County, Virginia government, www.fairfaxcounty.gov/.)*

Figure 1-4 Natural light. *(Lawrence Berkeley National Laboratories, http://www.capitolcommission.idaho.gov/images/log_natural_light.jpg.)*

FIGURE 1-5 Natural light in a public building. *(http://www.ecy.wa.gov/programs/swfa/greenbuilding/images/RecycledCarpet.jpg.)*

Use Appropriate Lighting

Using natural light is the best and most cost-effective solution for a home. Natural light is the best light for doing work or reading. If natural light is unavailable, proper placement and size of your lighting fixtures are critical. Appropriate lighting also will provide a better quality of room appearance. Costs are affected when multiple lights are used when only a single light is needed.

Use an Outdoor Clothes Dryer

Use an outdoor clothes dryer all year round. Figure 1-6 shows an excellent example of a $70 outdoor clothes dryer. This outdoor dryer was purchased in 2006 and has saved the owner over $2,000 to date. Made of aluminum and nylon cord, this dryer will last many years to come. The dryer also can be easily removed from the ground, folded, and stored.

In most parts of the world, many people hang their wet wash from a clothes line. Despite misconceptions, you do not need sunny days to dry your wash outside. During winter months, your clothing will dry rapidly because there is very little moisture in the cold winter air. Your clothing will allow the moisture to move quickly to the environment. If the temperature is below freezing, some of the moisture will remain and could freeze your clothing. This is the only occasion on which you would require the sun to complete the drying of your clothing.

FIGURE 1-6 Outdoor clothes dryer.

Use an Indoor Clothes Dryer

An indoor option is a simple clothing carousel or dryer. These devices do not cost much and last forever. Best of all, they require no energy. A full washload of jeans will take about 1 to 1.5 hours to dry in an electric or gas smart dryer. If you place the same jeans outside in almost any weather, most of the moisture is removed. Then you can put the same jeans in the indoor dryer and they will dry completely in 10 minutes, saving you 60 to 80 percent of your electric and gas bills. Figure 1-7 shows an indoor dryer. This unit is especially effective during the winter months.

Insulate Your Pipes

If immediate hot water and cold water are required at the faucet when you turn the faucet on, there is only one way to accomplish this—with pipe insulation. Figure 1-8 shows typical hot-water pipe insulation.

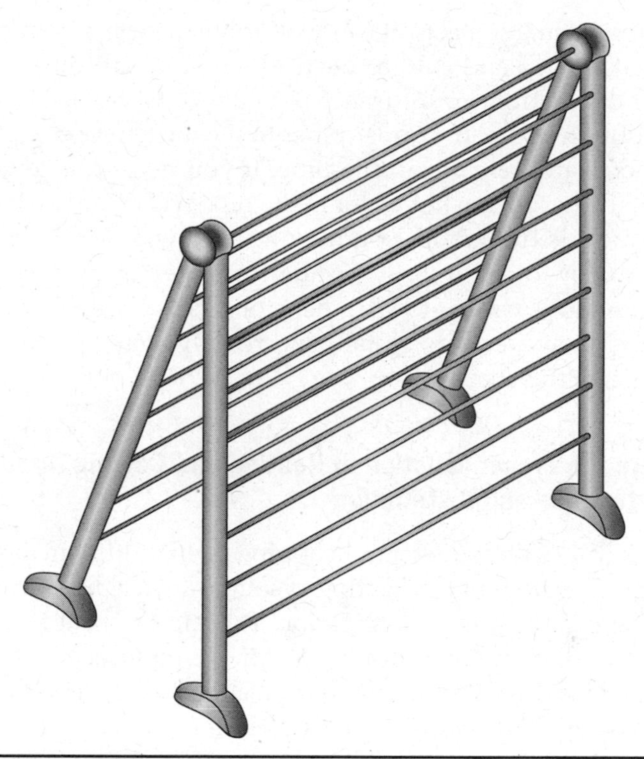

Figure 1-7 Indoor clothes dryer. *(United States Patent and Trademark Office, www.uspto.gov/.)*

Figure 1-8 Pipe insulation.

Pipe insulation acts like any other insulation; it prevents the transfer of heat to cold and cold to heat. When you turn on the hot water, that water comes from your furnace or your hot-water heater. The water travels through plastic or copper pipes to the faucet. After you turn the faucet off, hot water remains in the pipe. If you do not have your pipes insulated, the heat in the hot water in the pipe will dissipate into the air. When the faucet is turned on, you now need to wait for the cool water in the lines to flow out and the hot water from the furnace to reach you. This is a tremendous waste of water, heat, oil/gas, or electricity. Pipe insulation is inexpensive, easy to install, and readily available. There is no excuse for this inefficiency.

Ensure the Proper Function of Heating and Cooling Through Maintenance and Distribution

Baseboard (water) or steam heat pipes will require insulation. The pipes of the heating system transfer the heating medium to a final destination. Having the heat lost anywhere except where you want it in the home is not efficient. Heat fins (Figure 1-9) dissipate the heat at specific areas in specific amounts. This allows for proper distribution of heat throughout your home. This is the same principle as central air conditioning. Transferring the heat to the radiation point and proper maintenance of the heating fins will allow for efficient heating.

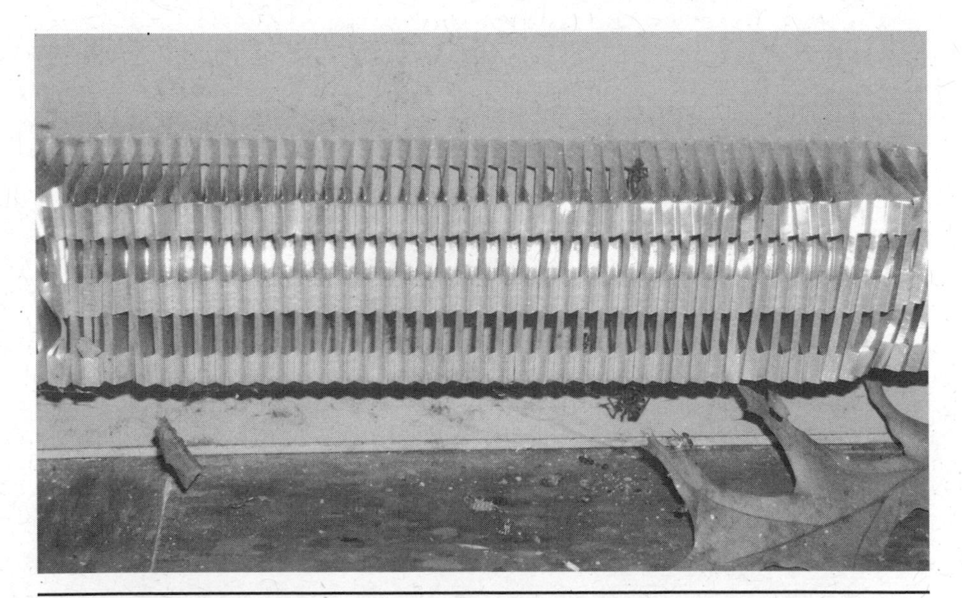

FIGURE 1-9 Heating.

Use Landscaping as Insulation

Create insulation for your home through the proper use of shrubs and trees. Wind is your enemy. Figure 1-10 shows how trees can be used to heat and cool a home. To create shade for cooling in the summer, use large trees. And you can slow the wind from scavenging heat from the foundation and first level during the winter with the use of shrubs.

FIGURE 1-10 Trees. *(U.S. Department of Energy, www.eere.energy.gov/.)*

If this is not reason enough to plant a tree, how about home value appreciation? In a typical suburban neighborhood, having two or more large trees, 50 ft tall or larger, can increase the value of your home by 15 to 20 percent. So plant a seed or transplant a tree and save 30 percent on your heating or cooling bill and increase the value of your property.

Identify Energy Problems

How do you know if your home can use a few improvements? Every home that I have ever visited for an audit could use a few improvements, and Figure 1-11 shows some of these. Figure 1-12 highlights the areas in a home that lose conditioned air. Before we get too specific, though, let's look at some simple indicators of poor home energy performance.

These are many indicators that will help you to identify common home problems.

Does your home:

- Feel cold or drafty?

FIGURE 1-11 Common home improvements. *(National Weather Service, www.srh.noaa.gov/.)*

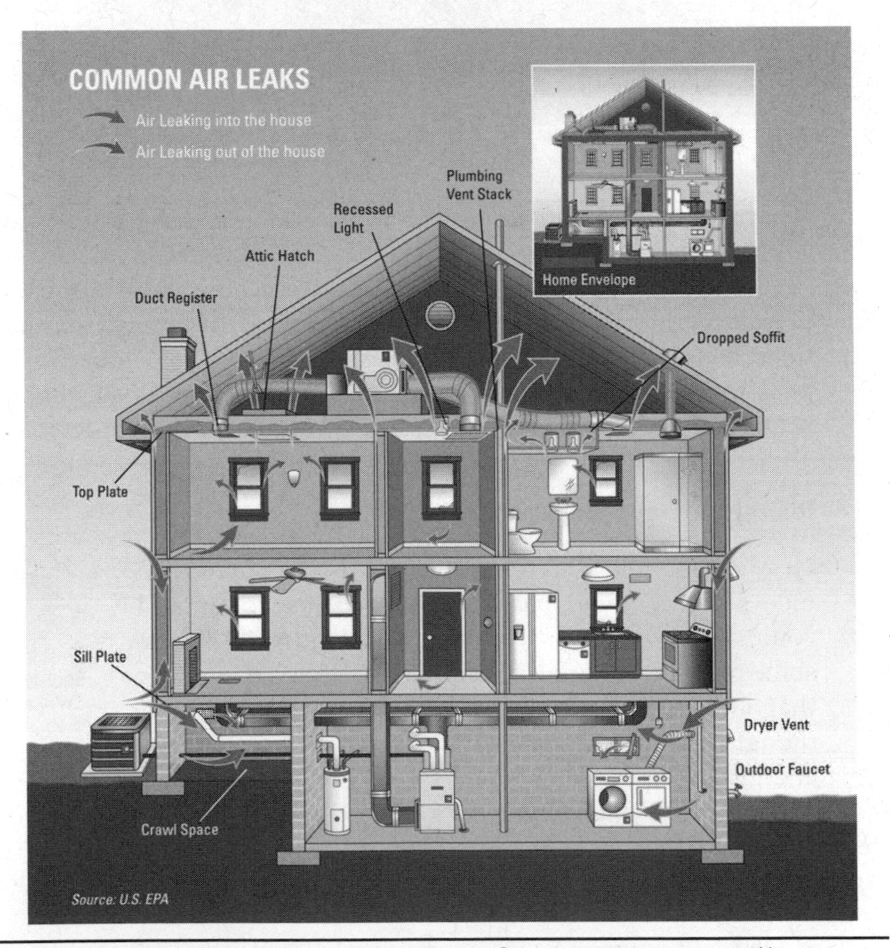

FIGURE 1-12 Common home air leaks. *(Energy Star, www.energystar.gov/.)*

- Have soaring energy bills?
- Have paint that is peeling even after a repair?
- Have excessive dust?

These are all signs that your home has great room for improvement. Addressing these problems will allow you to live more comfortably and save significant amounts of your after-tax income.

Let's review the most significant indicators of problems in your home and their causes.

Problem: High Energy Bills

Causes: High utility bills, exaggerated in the most extreme seasons (i.e., summer or winter), demonstrate the degree to which your home's energy envelope is incomplete. The most common violator in your home's energy envelope is insufficient attic insulation. The second most common cause is poor heating or cooling equipment. The final major offender is poorly performing windows and doors.

Problem: Drafty or Hot and Cold Rooms

Causes: Drafty rooms are often caused by leaks in the doors and windows, foundation, and attic. Hot and cold rooms are caused by inadequate insulation in the floors or walls. Poor heating and cooling from your heating and/or cooling system or poor duct performance from your system will create these problems.

Problem: Dry Air During the Winter Months When the Furnace Is in Use

Causes: Air is allowed to escape, usually through the attic. Warm air and moisture are allowed to flow from your home to the outside. This flow allows cool, dry air to flood in from the outside. The effect will be amplified as the cycle is repeated. As the heat and moisture leave the home, most people will turn the heat to a higher level and repeat the cycle. This is a synergistic waste of energy and money.

Problem: Mold, Mildew, Damp Basement, Moisture on Windows or Between Windows, and Peeling Paint

Causes: Water leaks from roof and walls from the outside to the inside. The problem also can result from internal pipe leaks. Water can migrate through walls if the exterior surface becomes porous because of lack of paint or siding. Inefficient insulation around windows or poorly performing windows will allow a temperature differential. This creates the opportunity for condensation. Moisture under paint can be attributed to a water leak, improperly applied paint, or improperly prepared painted surface.

Problem: Excessive Dust

Causes: Primarily, poorly sealed heating and cooling ducts. A poorly performing or poorly maintained heating or cooling system can create excessive dust. I should make one distinction here. Many heating systems expel black soot, not dust. The black soot is from combustion and is carcinogenic. If the dust is black, have your heating system repaired immediately. Clean the soot with a damp cloth, not a duster. The duster will only disburse the soot into the air, where it can be inhaled.

Problem: Ice Dams

Causes: Warm air leaks from the home into the attic and heats the underside of the roof. Snow and ice melt on the roof above. Water then accumulates in one area and forms a dam of ice and icicles (Figure 1-13).

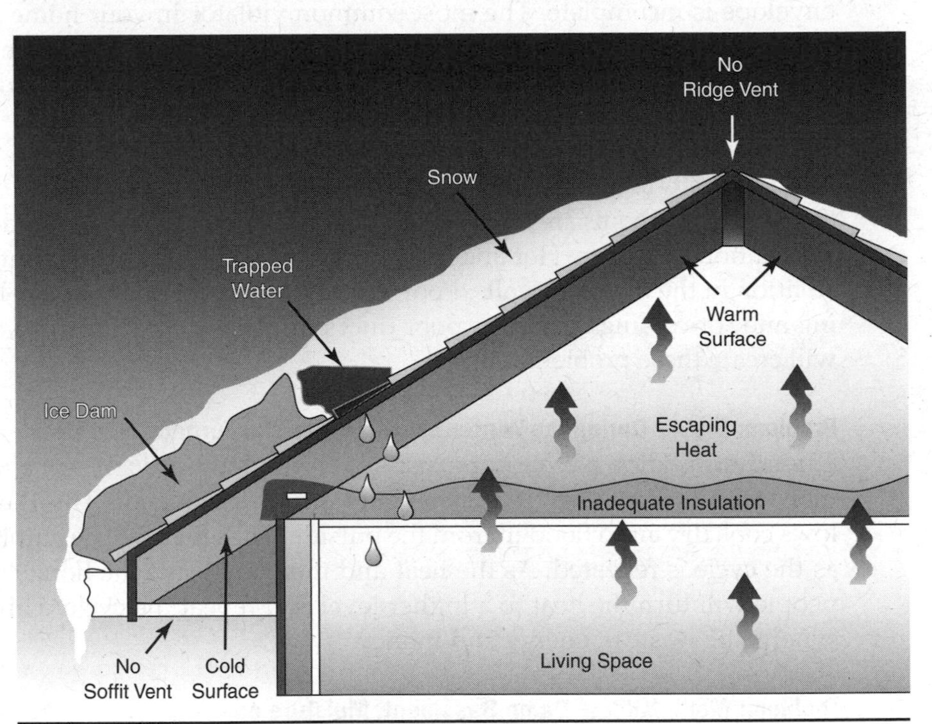

Figure 1-13 Ice dam. *(U.S. National Oceanic and Atmospheric Administration, www.noaa.gov/.)*

How Energy Is Transferred

I have clarified the most common home problems. Let's attempt to understand how we lose energy. We lose energy through different types of transference. You may not remember this, but your sixth grade child will know this material. Homeowners lose energy in three different ways.

The first type of transference is *conduction*, which is the transfer of energy through a medium. Metal is a good conductor. All types of metal will conduct heat and radiate its energy quickly. A metal pot in which water is boiling is a perfect example of conduction (Figure 1-14).

FIGURE 1-14 Conduction. *(State of California, www.energyquest.ca.gov/.)*

The second type of transfer is *convection*. Convection is the transfer of heat by the movement of air (Figure 1-15). A good example of this is your heating system. The air in the room flows over the radiator or the baseboard heating units. As the room's air contacts the hot surface of the radiator or heating unit, that surface releases heat to the cooler air.

Radiation is the third and final method of energy transfer that concerns our homes. The best and most common form of radiant energy is sunlight. Electromagnetic waves transfer energy to the objects they encounter. A good example is when sunlight hits your floor. The floor becomes warm over time. Radiant energy is solar radiation energy emitted by the sun and absorbed by the earth (Figure 1-16).

Why is our sixth grade science information important? This is how our homes lose energy. If you understand these simple concepts, then you can help to prevent the loss of energy and therefore the loss of your income.

Where We Proceed from Here

What does all this mean for you? Traditionally, we have accepted the inefficiencies in our homes. Poor construction, little or no insulation, and

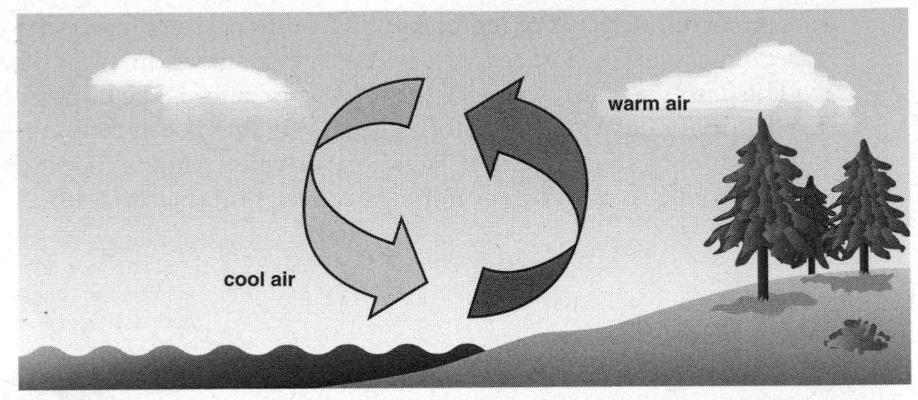

Figure 1-15 Convection. *(State of California, www.energyquest.ca.gov/.)*

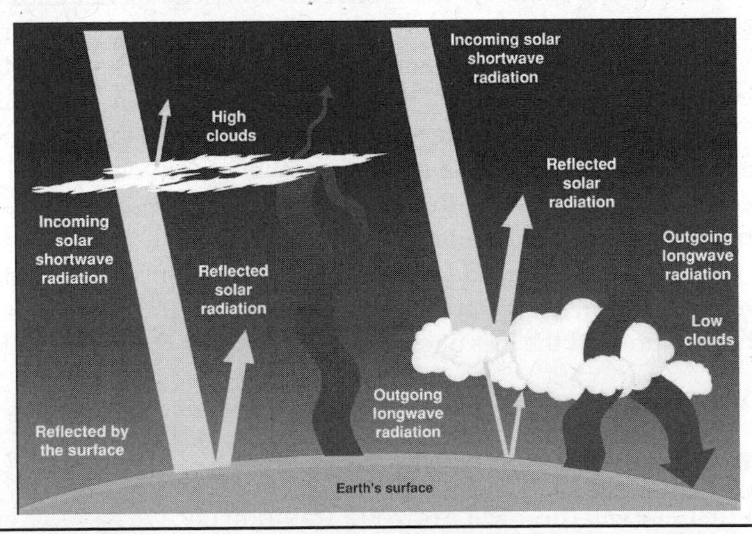

Figure 1-16 Radiation. *(http://veimages.gsfc.nasa.gov/103/Clouds_effects.jpg.)*

lack of planning for heating and cooling reflect the approach to home building during the last century. I do not recommend that you demolish your home or replace all the energy-using appliances in your home. What I do recommend is that you make a few habit changes and replace or upgrade items in your home as they need to be replaced.

Demolishing your home and building a completely energy-efficient home may sound like a good idea. However, this solution is not economically or environmentally friendly. Owning a new home that costs $300 a year for utilities is a terrific idea. Having the old home carted away to the town dump is not earth friendly, however. If you perform all the low- and no-cost solutions presented in this book, you will find that you may save as much as 50 percent of your energy usage. I would be willing to wager that if you put the correct amount of insulation in your home, you would see a savings of 30 percent at minimum.

Energy efficiency and living better are possible for very little cost. Saving energy will increase personal income and reduce greenhouse gases dramatically. Homeowners can improve both their personal and global footprint because they are one and the same.

Continue to the next chapter. Chapter 2 is dedicated exclusively to *doers*. I will help you to increase your home's efficiency immediately with items that are already in your home and therefore cost nothing. The entire chapter is dedicated to taking action.

CHAPTER 2

Home Improvements You Can Do Today at No Cost to You

You have purchased, borrowed, or stolen this book so that you could improve your home, save money, and therefore live better and more comfortably. That is a good first step. Let's begin now with some things that you can do immediately and at no cost to you. These changes may seem obvious, yet most people never consider how easy it is to save money. Most people never reflect on how to use resources. If you realize that every time you turn up the thermostat, turn on the water, or turn on a light, you are using a resource, it also follows that you are spending your hard-earned after-tax income on all of those resources. I will show you savings that you can achieve today with your refrigerator and water.

You and Your Refrigerator

Why do I start this chapter with your refrigerator? Because your refrigerator is like your heart. You buy a new refrigerator; plug it in; stuff it full of mayonnaise, hot dogs, hamburgers, and bacon; and it runs continuously and flawlessly for years. You never put another thought into your refrigerator until it stops working. The same could be said for your heart.

Your refrigerator is one appliance in your home that is used 24 hours a day, 7 days a week for many years continuously. Your refrigerator has an electric motor and a compressor that function continuously. Both these devices require a significant amount of energy to function. Second, and most important, have you ever completed any maintenance on your refrigerator? If you are like most people, the answer is no. So be kind to your heart and your refrigerator, and apply the lessons of this chapter to your refrigerator once a year. Let's get started.

Adjust Your Refrigerator Temperature

Your refrigerator's temperature should stay at between 35 and 41°F (~5°C) or less (Figure 2-1). A temperature of 41°F (5°C) is important because it slows the growth of most bacteria. Most people believe that bacteria die in a refrigerator, but this is not true. The cooler temperatures will not kill bacteria but instead will keep the bacteria from multiplying rapidly. Bacteria continue to grow; your refrigerator only slows the process. This is why food stays fresher in your refrigerator, preventing you from becoming ill.

Freezing at 0°F (–18°C) or less stops bacterial growth, although it won't kill all bacteria already present. All food has bacteria present. Your food does not go bad in the freezer, but it can suffer *freezer burn*. Freezer burn doesn't make food unsafe; instead, it affects the texture and taste of the food. Freezer burn occurs when frozen food has become damaged by ice crystals because of air reaching the frozen food.

Freezing causes an almost total block on the growth of microbes at –10°F and a total block at –18°F. Most bacteria can't live in temperatures that cold. Bacteria that cause harm to people grow at between 40 and 140°F. Once the food has defrosted, that's when the bacteria will start to grow again. I still would suggest to get rid of the frozen soup you have been saving for months because you won't be able to enjoy its taste anymore. Sorry for all the work and money you put into it.

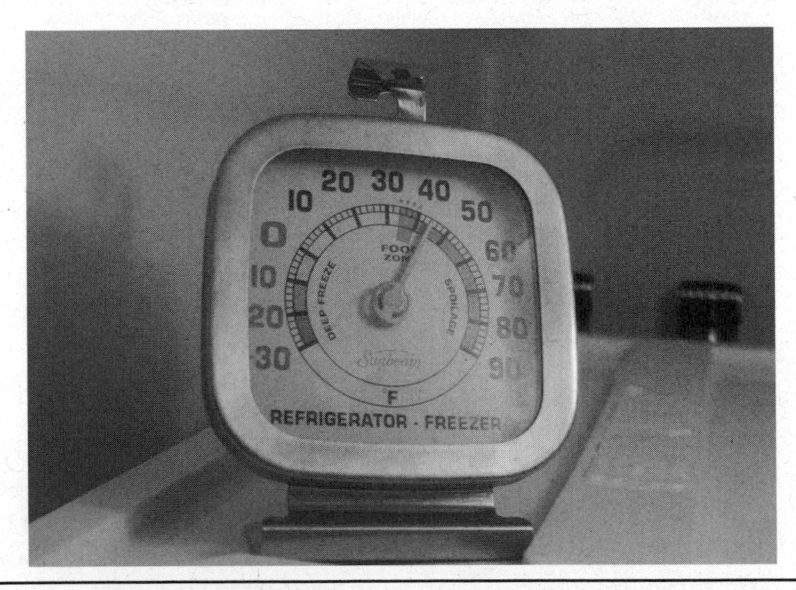

FIGURE 2-1 Refrigerator thermometer.

Clean the Seal/Door Gasket on Your Refrigerator Door

That seal between your refrigerator and its door (Figure 2-2) needs to be cleaned periodically. Soap and water and a damp sponge are all the tools you'll need. Be sure to clean the surface that mates with the seal on your refrigerator door (Figure 2-3) and between the accordion-style areas of the seal. Doing this will allow for a firm seal and no loss of cooling and therefore less use of electricity.

FIGURE 2-2 Refrigerator door seal.

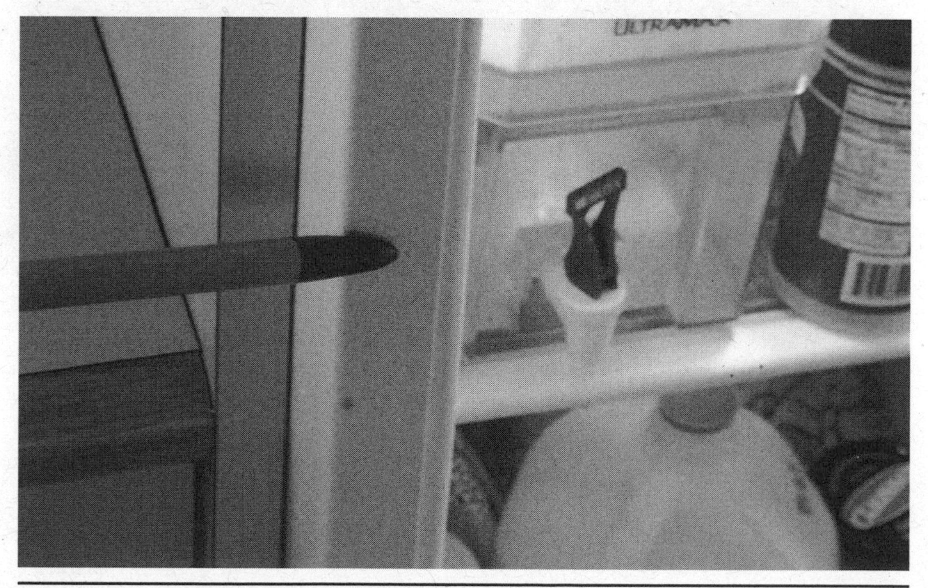

FIGURE 2-3 Refrigerator door seal mount.

Evenly Distribute Your Food Storage

Keep your refrigerator full but not cluttered (Figure 2-4). You want your refrigerator to be at an equal temperature in all areas. Have you ever taken items from the back of your refrigerator only to find them frozen? This phenomenon occurs because you have placed too many items too close to the cool-air inlet in your refrigerator. This also means that quite possibly the items on the door or in the front of your refrigerator may not be chilled to healthy temperatures. This also means that your refrigerator has been overworked and wasting electricity.

FIGURE 2-4 Evenly distributed storage.

Vacuum Your Refrigerator Coils

Most refrigerators are mounted on wheels, and there is a reason for this. You need to pull your refrigerator straight out from the wall to vacuum the coolant coils (Figure 2-5). Your refrigerator is very much like your air-conditioning system. You are either cooling your food in a box or cooling yourself in a box.

Figure 2-5 Refrigerator coils.

In essence, when a gas expands, it cools, and when a gas is compressed, it becomes hot. When you hear your refrigerator turn on, that is an electric motor starting a compressor. The compressor does exactly what its name implies: It compresses the "coolant" in the coils outside your refrigerator. If you touch the coils on the back of your refrigerator, they are hot or warm. The compressed coolant then is released through an expansion valve. This valve, again appropriately named, allows the coolant to expand and become cooler. This expansion occurs within your refrigerator, cooling the air and therefore the products in your refrigerator. The gas is then returned to the compressor to start the cycle again.

Occupy Unused Space in the Refrigerator or Freezer

You lose up to 30 percent of the cold air inside your refrigerator when you open the door. If you do not have many items in your refrigerator, place an appropriately sized closed food storage container in the empty place. Now when you open the door, the cool air in the food storage container remains in the refrigerator (Figure 2-6).

FIGURE 2-6 Rubbermaid food storage container.

Open the Refrigerator Less

Solution #16

Open the refrigerator door less and/or plan when you open your refrigerator. This may sound silly, but it can save you a great deal of energy. Most people never consider what they are doing. Know what items you need to prepare, take them from the refrigerator all at the same time, and return them to the refrigerator all at the same time. This is simple but very effective.

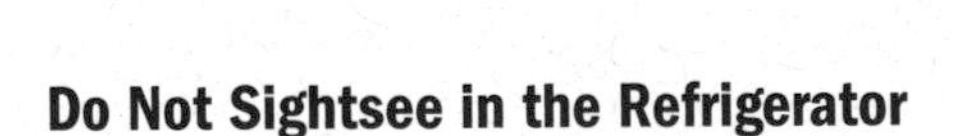

Do Not Sightsee in the Refrigerator

Do not sightsee. Everyone does it, but do not stand in front of the open refrigerator and make decisions. Go stare mindlessly into a cupboard or the pantry—it's more cost-effective.

Use the Water and/or Ice Dispenser

If you have a water and/or ice dispenser, use it. You lose 30 percent of the cool air during a simple door opening. An external ice or water dispenser is not a luxury but a money-saving device; use it.

Recycle Your Second Refrigerator or Freezer

If you have a second refrigerator or freezer, give it to someone who needs it. You are wasting your money. The supermarkets do a fine job of storing your food for you until you are ready to buy and use that food. If you have a second refrigerator or freezer so that you can buy items on sale and save money, you have been misled. Buying in bulk for paper towels is terrific, but not for refrigeration or freezer items. The money you save in sale prices is used by your freezer or refrigerator. In addition to the cost of storing chilled food, you have now frozen that fresh meat into a block of unidentifiable gray matter. We will do the calculations together later in this book. By the end of this book, you will be able to do your own calculations.

The Importance of Fresh Water

Now that your refrigerator is clean and operating correctly, let's flow to the next subject—water. Water is the most precious substance on earth, yet water has been and continues to be the most wasted resource that we have. Water in the developed world traditionally has been cheap and accessible. It is only now that problems are occurring with droughts and pollution that water is beginning to be treated as a precious resource.

The best and most cost-effective way to reduce your water consumption is to examine and modify your habits (Figure 2-7). Some of the sim-

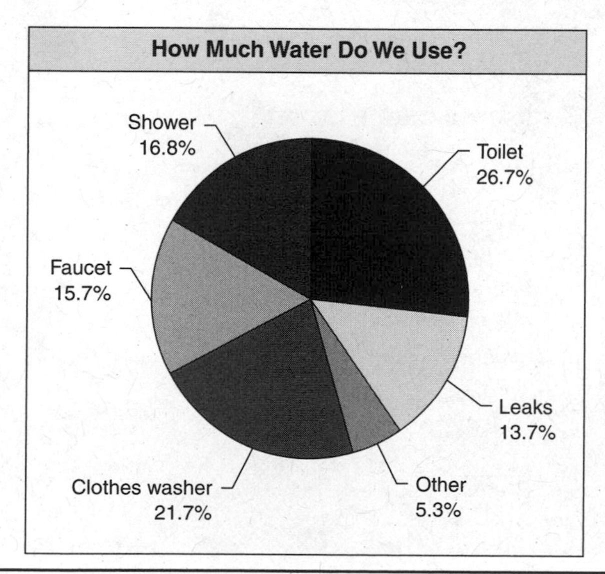

FIGURE 2-7 Water usage chart.

plest and least expensive ways to conserve water involve making small changes in how you use water. Most people do not think of water as a resource because it has been cheap and plentiful. I will demonstrate how you can reduce your current water consumption habits and water, sewer, and energy costs by modifying your behavior.

Use Your Home Water Meter to Check for Hidden Water Leaks

Solution #20

First, let's identify if you have a problem. You have a built-in diagnostic tool—your water meter (Figure 2-8).

Read the house water meter, and then wait for a 2-hour period when no water is being used. If the meter does not read exactly the same, there is a leak. Next, you must identify where you may have one or more leaks. Interior pipes that leak are easy to find; water stains will indicate potential areas. More frequently, you will find leaking faucets, toilets, or appliances.

These are the areas in your home where you can reduce your water usage at no cost to you, saving you water, energy, and money. Remember, hot water is significantly more expensive to use than cold water. Reduce

FIGURE 2-8 Water meter.

your hot-water usage and you reduce your energy bill by the same percentage. The bathroom can account for about 75 percent of the water used inside the home. Here are some tips for reducing your expenses without spending any money or changing your life.

The Bathroom

Take Showers Instead of Baths

Showers will save four times the water or more of baths, depending on the length of the shower.

Take Shorter Showers

A 4-minute shower uses approximately 30 gallons of water, and long, hot showers can use 5 to 10 gallons of water every minute.

Do Not Warm Up the Shower

Do not run the shower for long periods of time before getting in. If you are waiting for the water to get warm or using the shower to heat the room, you are using the wrong tool to fix a separate problem. Fix your plumbing or heating problem. Do not use your shower as a heater. That is very inefficient.

Turn Off Water While Lathering

Consider bathing small children together. If your shower has a single-handle control or shut-off valve, turn off the flow while soaping or shampooing.

Check the Shower Diverter Valve

The valve in the tub that you turn or pull to allow water to flow to the showerhead is the *diverter valve*. This valve is notorious for leaking. If this valve is leaking, your hot water is bypassing your body and going directly down the drain. You may be bypassing one-quarter of the water that should be coming from your showerhead. If the valve is leaking, it should be replaced.

Turn Down Water to Adjust Temperature

When adjusting water temperature, instead of turning the water flow up, try turning it down. If the water is too hot or cold, turn the offender down rather than increasing water flow to balance the temperatures.

Check Your Toilet for Leaks

Check your toilet for leaks by putting a little food coloring in your toilet tank. If the color begins to appear in the bowl without flushing within 30 minutes, you have a leak, and the bowl should be repaired immediately. Most replacement parts are inexpensive and easy to install. A leaking toilet can lose as much as 200 gallons of water each day.

Don't use the toilet as a wastebasket. Avoid flushing the toilet unnecessarily. Dispose of tissues, cotton balls, and other such waste in the trash rather than in the toilet. Not only is flushing these items clogging your septic system, they are also polluting the environment.

Reduce the Volume of the Toilet Tank

Put a brick or plastic bottles in your toilet tank (Figure 2-9) to reduce the amount of water used with each flush. By putting an object inside the toilet tank, you displace the holding capacity of the tank. Older toilets use 5 or 10 gallons per flush. Keep the external item safely away from the operating mechanisms in the tank. Modern toilets use 3 gallons or less.

Clean Your Razor in the Basin, Not with Running Water

Rinse your razor in the sink. Fill the basin with a few inches of warm water. This will rinse your razor just as well as running water with far less waste of water.

Turn Off Water While Brushing Your Teeth

Turn off the water after you wet your toothbrush. There is no need to keep the water running while you brush your teeth. Just wet your brush and use a glass for rinsing.

Brush your teeth first while waiting for the water to get hot, and then wash or shave after filling the basin.

FIGURE 2-9 Toilet tank.

Never thought of this, did you? What I am showing here is that with a little attention to how you move through life, life itself can be better for you.

The Kitchen

Approximately 8 percent of in-home water use takes place in the kitchen.

Don't Rinse Dishes Before Loading the Dishwasher

You can save up to 20 gallons of water per load. If you need to prewash your dishes, it is time to buy a modern, efficient dishwasher. Most makers of dishwashing soap recommend not prerinsing dishes.

Wash Only Full Loads in the Dishwasher

An efficient dishwasher usually uses much less water than washing dishes by hand. Operate automatic dishwashers only when they are fully loaded, or properly set the water level for the size of load you are using. If your dishwasher is noisy or uses a large amount of water or electricity, it needs to be replaced.

I am always told, "I do not have the money to replace my old appliance." This is not true. You do have the money; you just do not know it. Most new, energy-efficient appliances will pay for themselves with savings within 5 years. If you replace your dishwasher with a new model and keep it for 7 years, you will have paid for the dishwasher in full and made a profit. The difficulty is that most people do not plan for major appliance replacement.

When Washing Dishes by Hand, Fill Both Basins or Use a Separate Rinse Container

When washing dishes by hand, fill one sink or basin with soapy water. Quickly rinse your dishes under a slow-moving stream from the faucet. If you have a double-basin, fill one with soapy water and one with rinse water. If you have a single-basin sink, gather washed dishes in a dish rack and rinse them with a spray device.

Store Cold Drinking Water in the Refrigerator

Store drinking water in the refrigerator rather than letting the tap run every time you want a cool glass of water. A water dispenser (Figure 2-10) is the best solution for this problem. If you do not have one, use a large container that will meet your family's needs.

Defrost Foods at Room Temperature

Do not use running water to thaw meat or other frozen foods. The only way you should defrost food is overnight in the refrigerator. Most foods, when taken from the freezer to the refrigerator, will require 24 hours to defrost for cooking. By placing the frozen item in your refrigerator, you are cooling your fridge and warming your frozen food. Using the defrost setting on your microwave is just a plain waste of energy. Why do you currently use this method for defrosting? Because of habit.

Figure 2-10 Water dispenser.

Do Not Use Waste Disposal Units

Kitchen sink disposal units require lots of water and electricity to operate properly. In place of your garbage disposable unit, begin a compost pile as an alternate method of disposing of food waste. Garbage disposals also can add 50 percent to the volume of solids in a septic tank, which can lead to malfunction and maintenance problems. A compost pile also will fit into your new lawn maintenance approach without chemicals or overwatering.

Rinse Fruits and Vegetables in a Bowl

Don't let the faucet run while you clean vegetables. Rinse your fruits or vegetables in a closed sink or a bowl of clean water.

Other Indoor Water Usage

Use Gray Water to Water Plants

Never let water go down the drain when there may be another use for it. Today, gray-water systems are used for watering house plants and for lawn and garden care. A gray-water system consists of a place to store your sink and bath water and a method to use it to water your lawn or plants. Your gray-water system can run directly to your plants, or the gray water can be stored in a tank and used at another time. Today, there are many options, and you can choose the option that best fits your needs.

Stop All Water Leaks

Fix any leaks that you have, especially hot-water leaks. *Remember, you are paying for each drip.* If the leak is hot water, you are paying for water and electricity or oil/gas, all of which are being completely wasted.

Repair dripping faucets by replacing washers. If your faucet is dripping at the rate of one drop per second, you can expect to waste 7,300 gallons per year, that is, 7.4 gallons a day. This small leak will add to the cost of water, sewer utilities, or your septic system.

Check Your Water Well Pump

Check your pump. If you have a well at your home, listen to see if the pump turns on and off while the water is not in use. If it does, you have a leak.

Wash Full Loads of Clothing

Use your clothes washer only for full loads. Clothes washers should be fully loaded for optimal water conservation. With clothes washers, avoid the permanent press cycle, which uses an added 20 liters (5 gallons) for the extra rinse.

Wash Clothing Only in Cold Water

Only wash clothing in cold water. Washing in hot water does not kill germs; this is a misunderstanding. Soap kills germs. If you have clothing that is stained, pretreat or let soak for hours in the cold water.

Outdoor Water Usage

Water Your Lawn Only When Required

Water your lawn only when required. A good way to see if your lawn needs watering is to step on the grass. If it springs back up when you move, it doesn't need water. If it stays flat, then the lawn requires watering.

Don't overwater your lawn; overwatering will allow for the growth of fungus that will kill and displace your grass. Deep soak your lawn. When watering the lawn, do water it long enough for the moisture to soak down to the roots, where it will do the most good. A light sprinkling can evaporate quickly and tends to encourage shallow root systems. Weeds prefer a short, light watering; your lawn requires a deep watering.

Plant Appropriate Indigenous Plants

Plant only drought-tolerant and indigenous grass. Of all the plants on your property, your lawn probably requires the most regular watering. When you must fill a bare spot, use drought-tolerant grass seed. Many "big box" stores or governments are offering rebates to use this grass. You should be able to obtain some for free.

Replace Your Grass with . . . Anything

If you are having difficulty with clover, moss, or other plants, why not use them as your lawn? Alternatives to traditional grass are becoming popular. Do not limit your options. Figure 2-11 shows a nice soft, green lawn of moss. A lawn of well-maintained clover appears very similar to a traditional lawn (Figure 2-12).

Do Not Water Your Lawn in Direct Sunlight

Water your lawn during the early parts of the day; avoid watering when it's sunny or windy. Watering in the early

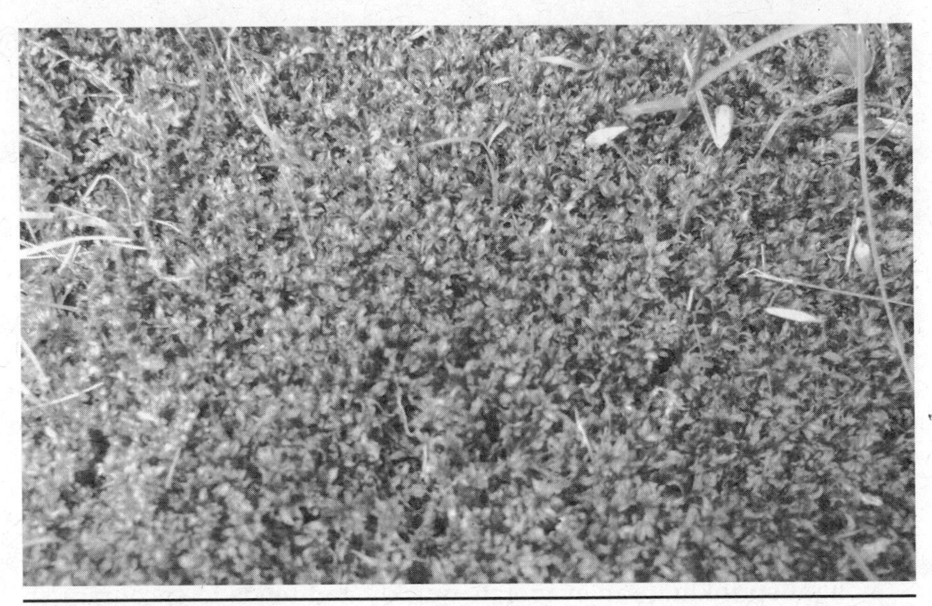

Figure 2-11 Moss lawn.

Figure 2-12 Clover lawn.

morning generally is better than at dusk because it helps to prevent the growth of fungus. Early watering (and late watering) also reduces water loss to evaporation. Watering early in the day is also the best defense against slugs and other garden pests.

Check Sprinklers for Proper Usage

Position your sprinklers so that the water will fall on the lawn and shrubs, not the paved areas. Install sprinklers that are the most efficient for each use. Microhoses and drip-irrigation and soaker hoses are examples of water-efficient methods of irrigation. Check your sprinkler system and timing device regularly to be sure that they are operating properly.

Use a Moisture Sensor for Automatic Sprinklers

If you have an automatic sprinkler system, be sure that it includes a moisture sensor. We have all seen homes where the sprinkler system was operating during a rain storm. Redundancy is good in airplanes but not in lawn watering.

Use Proper Nozzles for Hand Watering and Distribution

If you water with a hose, outfit your hose with a shutoff nozzle that can be adjusted down to fine spray so that water flows only as needed. When finished watering, turn the hose off at the faucet instead of at the nozzle to avoid leaks.

Do Not Leave Sprinklers Unattended

Do not leave sprinklers or hoses unattended. Your garden hoses can pour out 600 gallons or more in only a few hours, so don't leave the hose or sprinkler running all day.

Use a Timer for Automatic Sprinklers

Automatic sprinkler systems frequently have built-in timers. Be sure that the timer is set properly. On systems without timers, if necessary, use a kitchen timer to remind you to turn the system off.

Design Your Landscaping to Use Water Runoff

Landscaping is a great way to design, install, and maintain both your plantings and irrigation system in a way that will save you time, money, and water. Design your landscaping to capture water and not allow it to run off into the storm drains. The water from your roof and pathways can be directed onto your property for immediate use. An alternative is to store the runoff water in your gray-water system. Figure 2-13 shows how to direct and use water runoff on a piece of property, and Figure 2-14 shows how water runoff from a pool deck can be used.

FIGURE 2-13 Use of water runoff in landscaping.

FIGURE 2-14 Use of water runoff from a pool deck.

Cut Your Lawn to the Proper Level

Raise the blade on your lawn mower to at least 3 inches above ground. A lawn regularly cut this high encourages the grass roots to grow deeper, shades the root system, and holds soil moisture better than a closely clipped lawn.

Leave Lawn Clippings on the Lawn

Use a mulching mower that will leave the clippings on the lawn. Lawn clippings are full of moisture and nitrogen, the two items your lawn requires to grow. Mulching mowers have four blades to reduce the size of the grass clippings (Figure 2-15).

FIGURE 2-15 Mulching mower blade.

Avoid Overfertilizing

Avoid overfertilizing your lawn. The application of fertilizers increases the need for water. If you understand lawn care, you know that there is no need for chemical

fertilizers or weed killers. Use the topsoil you have created with your new mulcher. Spread the topsoil lightly on top of your current lawn. This will add the necessary nutrients and moisture that your lawn needs.

Remove Weeds by Hand

Remove weeds by hand. I know what you are going to say, but have you ever tried it? I found that I could weed a third of an acre of my lawn in about 15 minutes. This is less time than it takes to retrieve and fill a spreader and disburse harmful chemicals all over my yard. In flower beds, you simply turn the weed over, and it becomes food for your plants. Use a small hand shovel to remove weeds from your lawn. Plunge the shovel in the ground under the weed's roots, pull out the weed, and place a handful of grass seed in its place. Shake off the topsoil from the roots of the weed to cover the grass seed. I found this method to be fast and effective, and I do not have to worry about poisoning myself or my family.

Use Mulch to Retain Water

Use mulch to retain moisture in the soil around your plants. Mulching also helps to control weeds that compete with plants for water. If you have an area in your yard where grass will not grow, put down a layer of mulch (Figure 2-16). This will make the area look neat, clean, and maintained and, over time, will encourage the growth of grass.

FIGURE 2-16 Proper use of mulch.

Group Plants Together Based on Water Requirements

Always plant native drought-tolerant grasses, ground covers, shrubs, and trees to reduce water usage. Once established, these plants will thrive and do not need to be watered as frequently. Native plants usually will survive a dry period without any watering. In addition, group plants together based on similar watering requirements.

Use a Rain Barrel to Collect Water

Add organic matter and use an efficient watering system for shrubs, flower beds, and lawns. Adding organic material to your soil will help to increase its absorption and water retention. Areas that are already planted can be dressed with compost or organic matter. You can greatly reduce the amount of water used for shrubs, flower beds, and lawns with strategic placement of soaker hoses, rain barrels (Figure 2-17), catchment systems, and simple drip-irrigation systems. A water meter can be easily added to your hose to monitor water usage. Avoid overwatering plants and shrubs because this actually can diminish plant health and cause yellowing of the leaves.

FIGURE 2-17 Rain barrel water-collection system.

Plant Shrubs That Will Help to Retain Water

Plant and grow drought-resistant lawns, shrubs, and plants. If you are planting a new lawn or overseeding an existing lawn, use drought-resistant grass seed. Replace perennial borders with native plants. Native plants will use less water and be more resistant to local plant diseases. Consider using a low-maintenance, drought-resistant yard design. Plant shrubbery staggered with plants that will retain water and help to reduce runoff.

Create and Use Your Own Mulch

Put a layer of mulch around trees and plants. Mulch will slow evaporation of moisture while discouraging weed growth. Adding 2 to 4 inches of organic material such as compost or bark mulch will increase the ability of the soil to retain moisture. Use a drip line to water, and press the mulch down around the drip line of each plant to form a slight depression, which will prevent or minimize water runoff.

Use a Broom to Clean in Place of a Hose

Use a broom, not a hose, to clean driveways and sidewalks. For thousands of years, brooms have worked perfectly fine. Brooms are inexpensive or free, last for years, and create no pollution.

Maintain All Watering Equipment

Check all hoses, connectors, and spigots regularly. Use hose washers between spigots and water hoses to eliminate leaks (Figure 2-18). Check for leaks, and replace washers as needed. A properly used and maintained hose can last your entire life. During the winter months when your hose is not in use, remove it and store it in a shed or garage. Also check for leaks in pipes, faucets, and couplings. Leaks outside the house may not seem as bad because they're not as visible, but they can be more wasteful than leaks inside your home. You can waste thousands of gallons outside without notice (Figure 2-19).

FIGURE 2-18 Typical water loss in a home hose water system.

FIGURE 2-19 Typical outside water loss in an average home.

Always Use a Hose with a Nozzle

Outfit your hose with a shutoff nozzle that can be adjusted down to fine spray so that water flows only as needed. Don't allow the hose to flow freely while washing your car.

Wash Your Car on the Lawn

If you can, wash your car on your lawn. Use a bucket of soapy water, and use the hose only for rinsing. This simple practice can save as much as 150 gallons when washing a car. Use a spray nozzle when rinsing for more efficient use of water.

Use a Commercial Car Wash

Consider using a commercial car wash that recycles water. Most modern car wash services recycle the water, and this is good. However, these services do use a great deal of energy.

Do Not Install Ornamental Water Features

Avoid the installation of ornamental water features such as fountains or waterfalls unless the water is recycled. If necessary, locate the feature where there are minimal water losses owing to evaporation and/or wind drift.

Use a Modern Pool Filter

If you have a swimming pool, consider using one of the new water-saving pool filters. Older types of pool filters use more water.

Use a Solar Pool Cover

If you heat your pool, use only a solar pool cover. A solar pool cover is a clear sheet of plastic that floats on top of your pool. Sunlight flows through the plastic into the water and warms the water. There are two benefits to heating your pool with solar energy. The first benefit is that the energy is free, and it requires a great deal of energy to heat the water in a pool. The second benefit is that the solar cover prevents the heat energy from leaving the pool and also prevents evaporation. On a very dry, hot day, I have noticed that I can lose ½ inch of water from my 36- by 16-foot pool. A solar cover can reduce evaporation dramatically.

Beyond your home, you can apply all these new habits universally. The U.S. government has an excellent Web site that addresses these issues (www.epa.gov/watersense/tips/index.htm). This Web site provides tips for consumers, businesses, utilities, and communities. Several conversion tools are available on the Internet that can be used to make your calculations easier. See the following sites:

www.onlineconversion.com/volume.htm
www.mathconnect.com/volumel.htm

If you are truly brave, you can provide your child with the drip calculator from www.awwa.org/awwa/waterwiser/dripcalc.cfm. Be forewarned, your child will want to test this theory over and over.

Now that we have completed an entire chapter of no-cost solutions, let's go shopping. For as little as a few hundred dollars today, you may be able to save thousands of dollars each and every year. Chapter 3 will discuss low-cost solutions for your home. Do not forget that there is a no-cost solution section in each chapter, so read on, and let's continue to improve your home.

CHAPTER 3

Low-Cost Solutions

There are hundreds of ways to reduce the energy consumption in your home. For most of us, the reason we do not save energy is that *change is difficult*. Increasing the heat during the winter months is an easier solution than insulating the attic. Increasing the heat also increases your monthly expenses, though. Insulating the attic will save you money this year and every year after. I recommend that you attempt these energy-efficient solutions as soon as possible, but it is easier to make improvements during temperate weather. You do not want to be installing insulation in the attic during the summer months or new windows during the winter. While this may seem obvious, these ambitious types of installations occur every day.

The second item to be addressed is that you must make these changes when you are ready. Do not make efficiency changes to the detriment of your finances or your health. Finally, address the issues that solve the biggest problems. Changing your lighting to compact fluorescent lighting is good, but if you are paying $7,500 each year to heat your home, you should choose to insulate your home first.

Address the Biggest Costs in Your Home First

The easy way to identify the biggest energy consumers in your home is by your bills. Money will equate directly with kilowatts or British thermal units (Btu's) used. Do not be concerned about understanding every aspect of energy. Focus on the biggest expenses. Reduce your home costs, and you will save energy. Table 3-1 shows average household energy usage. This energy consumption chart should provide you with a place in your home to begin reducing your costs and—remember therefore—increasing your income.

Table 3-1 Average domestic energy consumption per household in temperate climates

Heating	12,000 kilowatthours/year (1,400 watts)
Hot water	3,000 kilowatthours/year (340 watts)
Cooling/refrigeration	1,200 kilowatthours/year (140 watts)
Lighting	1,200 kilowatthours/year (140 watts)
Washing and drying	1,000 kilowatthours/year (110 watts)
Cooking	1,000 kilowatthours/year (110 watts)
Miscellaneous electric load	600 kilowatthours/year (70 watts)

Energy Costs

If you would like to see how much each appliance in your home is costing you, try this very cool calculator: http://odograph.com/experiments/php/instantkwh.php. This is true not only for your lighting but for all your appliances. Electricity and water (Figure 3-7) are the two items that no home can live without: www.mde.maryland.gov/Programs/WaterPrograms.

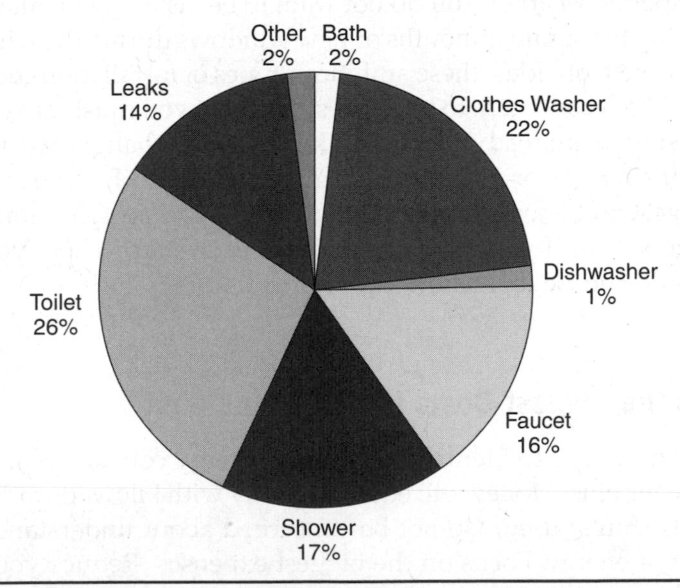

Figure 3-1 Water usage chart. *(http://en.wikipedia.org/wiki/Domestic_energy_consumption.)*

Let's take a short break and accomplish some immediate cost savings.

Lighting

Let's begin with the simple and most effective solutions.

Turn Off Lights When Not in Use

Turn off the lights (Figure 3-2)! Does it get any easier? It is always more efficient to turn lights off when you do not need them. Even turn off your new energy-efficient lights when you do not need them. This rule applies to all electronics.

FIGURE 3-2 Light switch.

Consumers feel good about making the transition from incandescent light bulbs to compact fluorescent lights (CFLs) or light-emitting diodes (LEDs). A new problem now arises: People who have changed to CFLs do not feel the urgency to turn off the lights because the lights are energy efficient. This is not right. When lights are not needed, turn them off. Your parents were correct, turn off the lights!

Early CFLs required a few seconds to a few moments to warm up and provide full light. During this time, the bulbs require additional energy. The rule of thumb was to leave a CFL on for 15 minutes or more. This is no longer the case. CFL bulbs warm up quickly now and require less energy to do so. Always turn off lights when not in use.

Use the Natural Light Currently Available in Your Home

Open your blinds and curtains (Figure 3-3). I will vary this rule in future chapters for heating and cooling purposes, but many people turn on lights when all that is required is to open the blinds. The sun is delivered free to your home every day. Why not let it in and use it? During the cooler months, the additional sunlight will help to heat your home. In the warmer months, shades can redirect the heat while still providing the home with ample lighting.

FIGURE 3-3 Natural lighting. *(www.michigan.gov/images/dmb/Mason_Garden_conf_rm_192154_7.jpg.)*

Use the Appropriate Light for the Appropriate Job

Use appropriate lighting when you must use electric light. If you are reading or doing paper work, the light source must be close to the material being viewed. Turning on overhead or room light is not sufficient. Using the appropriate lighting will save money and your vision.

Correct placement for room lighting often will eliminate the need for additional lights and therefore save energy. Check with the lighting manufacturer or reference the type of lights being used in your home. For example, recessed lights are designed to illuminate a certain square footage, with definite spacing between lights.

Use Motion Detectors

Solution #73

Use motion detectors to turn on lights and, more important, to turn off lights. Motion detectors are especially useful for outdoor lighting applications and children's rooms. Anyplace that hands-free lighting is helpful, motion lighting can be installed. Motion sensors placed in a child's room are especially useful, but it is important to note that children do not live only in their rooms.

Replace Incandescent Bulbs with CFLs

Floor lights, recessed lights, and even chandeliers all now can use CFLs. Yes, you can obtain CFLs at low cost in white, true light, or colors (Figure 3-4). You can find a CFL for any application, indoor or outdoor. The cost of these bulbs is the same as or a little more than a traditional incandescent bulb. However, a CFL lasts seven times longer; therefore, the cost of each bulb is actually less than that of an incandescent bulb. Tax

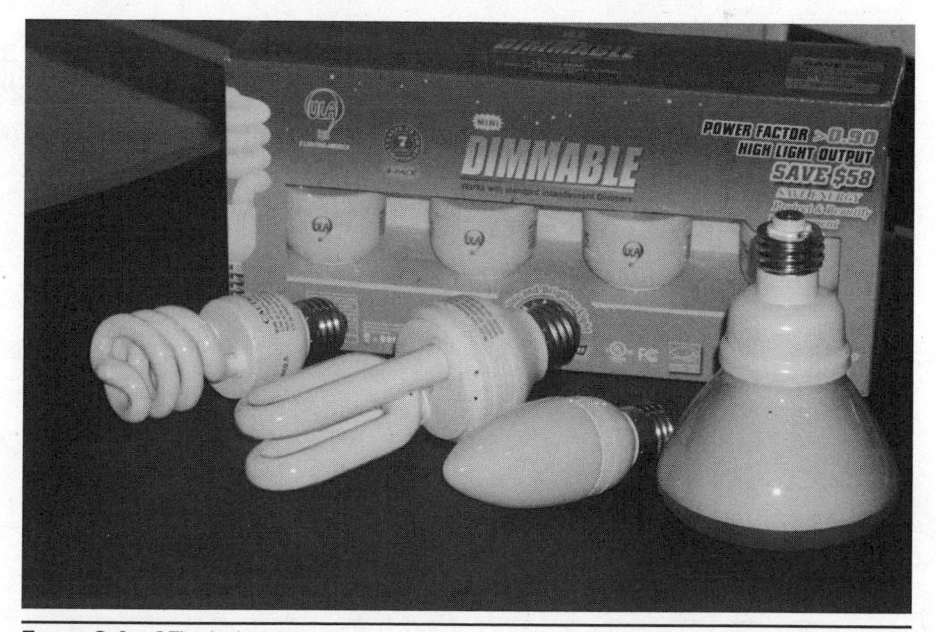

FIGURE 3-4 CFL choices.

credits and rebates can provide you with a profit for buying and using these bulbs. Many local utilities offer these bulbs for reduced prices or for free. In a kitchen with five recessed lights with 120-watt bulbs, you use a total of 600 watts. Replace these bulbs with five dimmable 23-watt CFLs, and you will provide the same function and illumination at a fraction of the energy cost. The five CFLs will use 115 watts; that is, five bulbs now use less energy than one of the old incandescent bulbs.

I read the following piece of advice all over the Internet. If you put CFLs in your home and you rent, then take the bulbs with you if you move. This makes sense. A better alternative is to speak to the landlord. If energy improvements can be made, maybe the landlord will pay for a portion or the full amount of the improvements. It does not require additional money to ask.

Keep Lights Away from the Thermostat

Solution #75

Keep lamps away from the thermostat. This can cause a false temperature reading at the thermostat and change your home temperature unbeknownst to you. While CFLs create less heat energy, they will create heat, so move or point lighting away from the thermostat.

Use Dimmer Controls

Solution #76

Use your dimmer controls for appropriate amounts of light. Modern dimmer switches allow for a nice variation of lighting themes. If full, bright light is not required, use the dimmer to reduce the lighting for the appropriate mood and energy savings (Figure 3-5).

FIGURE 3-5 Dimmer switches.

Use Timers for Your Lights

Use timers for appropriate operation of lights (Figure 3-6). Consumers are appreciative of convenience. Use timers to turn off all lighting and appliances when not in use. Have timers turn on lighting and devices prior to use. This can save as much as 15 percent of an average consumer's electric bill.

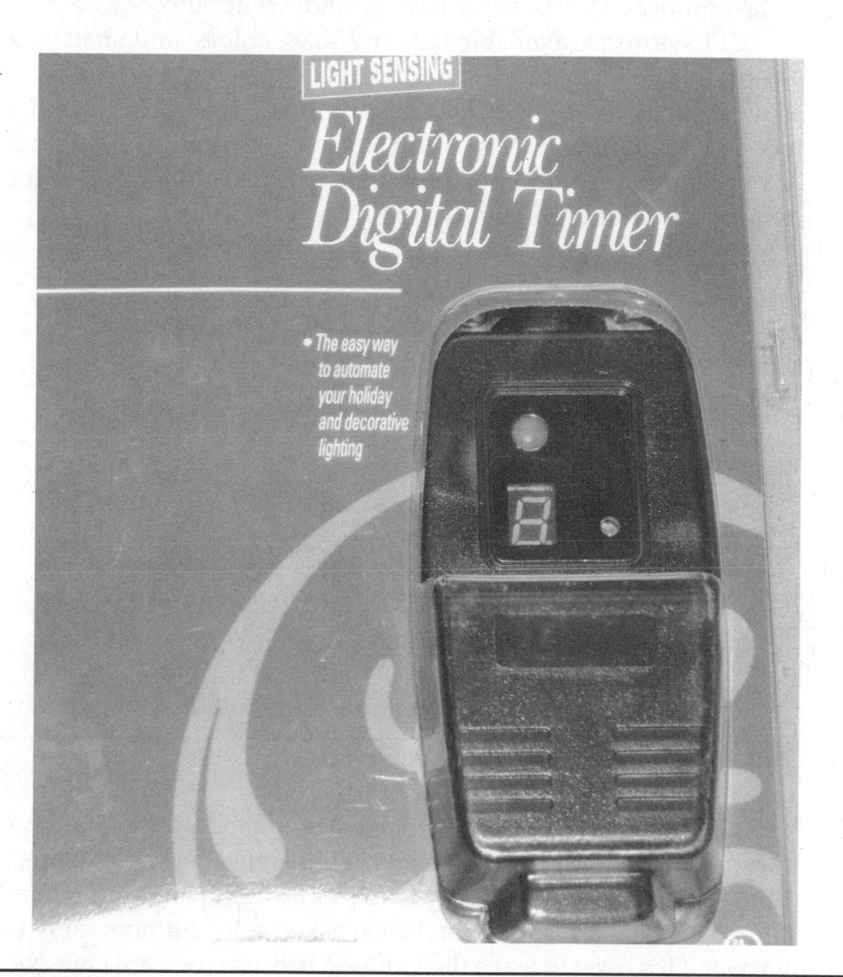

FIGURE 3-6 Timer.

Compact Fluorescent Lights (CFLs): A Few Facts

This section will provide you with additional information about CFLs. A typical CFL saves 75 percent or more of the electricity used by an incandescent light bulb. CFLs produce the same amount of light but less energy-wasting heat. A CFL bulb will last approximately seven times longer than an incandescent bulb. Recently, the cost of a CFL bulb has been reduced dramatically. Most of the CFL bulbs purchased during the past year have cost consumers approximately $1 per bulb because of the extended bulb life, the production of less heat and thus the use of less energy, and tax credits and rebates available. These factors make the use of CFL bulbs considerably less expensive than incandescent bulbs.

CFLs now are available in many sizes, colors, and shades, as well as decorative and dimmable styles (Figure 3-7). Today, you can replace every bulb in your home with a CFL bulb. Just a few years ago, CFL bulbs needed a few moments to warm up and provide full light. Today, most CFL bulbs come to full power within seconds. No longer do these bulbs require significant energy to start.

FIGURE 3-7 Compact fluorescent lights (CFLs).

Interestingly, CFLs have led to a new phenomenon that I call the *lazy theory*. Once the CFL bulbs have been installed in a home, it seems that homeowners have a tendency to leave the lights on more often than they did prior to CFL installation. When asked why, homeowners frequently answer, "It's okay to leave the lights on because they are energy-efficient." This is incorrect. Just because CFL bulbs are more efficient does not mean that lights should be left on indefinitely. As with any lighting, CFLs are most efficient when turned off. Conservation is always the best policy. When asked, "Should I leave lights on?" the answer is always "No." Always turn off lights when not needed.

From another perspective, it is important to dispose of burned-out CFLs properly. CFLs contain mercury, so it is necessary to contact your local government to ask about disposal and recycling programs. In addition, if a CFL breaks, leave the area for a few moments, and be careful not to inhale the dust. Return 10 to 15 minutes later and pick up the glass. Spray the area with a light mist of water or cleaner. Wipe up the mess, and repeat until all traces of the dust are gone. Place all cleaning materials and the pieces of the CFL in a sealed container and recycle. Wash your hands when done. Do not use a duster or a dry paper towel to clean up a broken CFL. This method can sweep the mercury dust into the air where it can be inhaled.

If Americans replaced all their incandescent light bulbs with CFLs, total electricity consumed would be reduced by 10 to 12 percent. Thus U.S. greenhouse gas emissions would be reduced by approximately 4 percent. Where electricity is produced from coal, each CFL will reduce CO_2 pollution by about 1,300 pounds over its lifetime. Coal produces an average of 10,000 British thermal units per pound, or approximately 1 kilowatthour. That's one 100-watt bulb turned on for 10 hours, or 1 kilowatt. If every household in the United States replaced just one incandescent light bulb with a CFL, the CO_2 pollution equivalent of 1 million cars would disappear.

The Future of Lighting

The future of lighting lies with the light-emitting diode (LED). LEDs have been used for years in your small electronic devices (Figure 3-8), but not for ambient lighting. This is the case for two reasons. First, LEDs were expensive in the beginning because they produce light through the movement of electrons through a semiconductor, and semiconductors were expensive to manufacture. And second, it was difficult at first to make a LED that produces white (or colorless) light.

Figure 3-8 Typical LED usage. *(www.publicdomainpictures.net/.)*

Incandescent bulbs create light by using an inexpensive metal filament. The filament becomes hot and glows with light. Incandescent bulbs are inefficient because most of the energy is converted to heat. CFLs generate light by passing electricity through a gas. The gas becomes excited and reacts with the fluorescent coating on the bulbs, producing visible light. While CFLs are significantly more efficient than incandescent bulbs, they still lose most of their energy (up to 80 percent) to heat.

LEDs produce light via the movement of electrons through a semiconductor. An LED loses only a very small portion of its energy to heat. Improvements in mass production of semiconductors are reducing the cost of LEDs, and new technology has brought the quality of the light produced closer to natural light (Figure 3-9).

FIGURE 3-9 Current-generation LEDs. *(www.publicdomainpictures.net/.)*

The benefits of LEDs are numerous and include:

- LEDs produce equal or greater luminance than existing lighting technology.
- The light from LEDs remains constant over time.
- LEDs now have excellent color quality.
- The light fixtures employing LEDs do not draw power on their own (i.e., a vampire load).
- LEDs have a long life span, approximately 25,000 hours of use.
- LED light does not flicker and turns on instantly.

LEDs are indeed the future of lighting. Do a simple Web search at a retailer such as Amazon.com, and you will find hundreds of options from which to choose.

LEDs belong to a class of lighting called *solid-state lighting*. Beyond today's LEDs, the next generation of lighting will include *organic light-emitting diodes* (OLEDs). These LEDs produce light when current is applied to a sheet of carbon-based compounds. The effect is to cause the carbon material to glow. These lights are not yet available commercially, but the technology is changing as rapidly as your computer. OLEDs can be almost as flat as a piece of paper. This technology will allow you to have low-cost lighting everywhere that you may need it.

The Department of Energy (DOE) estimates that LEDs will become the dominant lighting technology over the next 20 years. This would translate into $265 billion in energy savings, or a reduction in the need for 40 power plants. This technology also could reduce total lighting electrical demand by approximately 30 percent by 2027.

More Energy Savings

Use a Programmable Thermostat

Buy and use a programmable thermostat for your heating and cooling needs (Figure 3-10). A programmable thermostat can provide you with warm mornings on cool winter days and energy savings while you are at work. New thermostats can be monitored or changed via the Internet. A programmable thermostat can cost as little as $25.

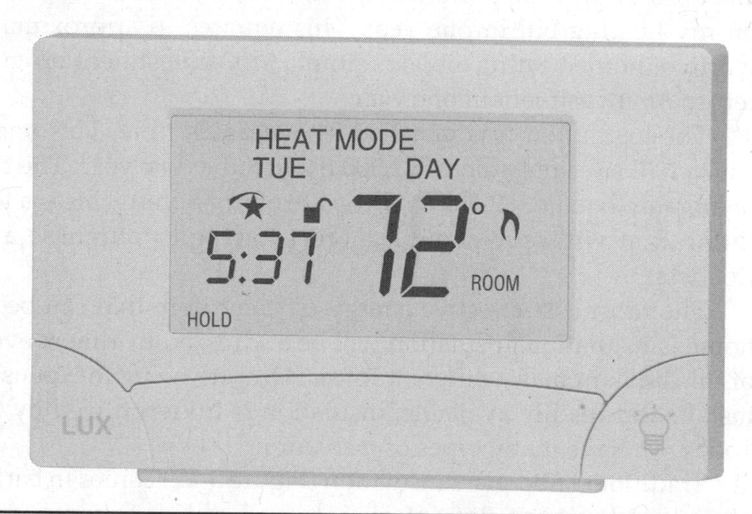

FIGURE 3-10 Programmable thermostat. *(www.energystar.gov/ia/partners/promotions/cool_change/images/Programmable_Thermo_4.jpg.)*

Reduce Vampire Load

A *vampire load* is a loss of power that is used to maintain devices in your home that are in ready or standby mode. The best example of a vampire-load appliance is your television. When the television is turned off, it still requires power—power to maintain the settings and power to respond to the remote control when you turn the set on for viewing. Products such as the GreenSwitch from http://greenswitch.tv can reduce your electric bill by 25 to 45 percent.

The GreenSwitch powers off all the connected light sockets in your home with one switch. This allows you to turn off all of your unwanted vampire loads. When power is required, again one switch allows power to flow to the devices. This is all accomplished with one master switch and remote electric sockets. A typical home can be outfitted with the GreenSwitch system for $500 to $1,000.

Add More Insulation

Having viewed and energy-audited many homes, I can say that I have never viewed a home built with the maximum amount of insulation required. *Every* home can use more insulation (Figure 3-11). For my own home I recently purchased $300 worth of insulation. I received a $100 tax rebate and a $100 rebate from the "big box" store where I bought the insulation. My total cost was $100. I placed the insulation in my attic and around my foundation sill. This $100 investment saved me 60 percent on my heating bill in one year. This equates to approximately $2,500 worth of home heating oil. This simple $100 investment provided a huge return on investment in one year.

The insulation was installed in one day's time. This one-time purchase will save me at least $2,500 each and every year. The total cost of ownership equals $82,500 over the life of my home. This is a low-cost solution that will save enough money for me to purchase a new Tesla Roadster.

The most cost-effective energy-efficient item that can be added to a home is insulation. Insulation will be addressed in almost every chapter of this book in many different forms. Despite being inexpensive, easy to install, and readily available, insulation is underutilized by consumers. Today we have many types of insulation.

Traditional fiberglass insulation (Figure 3-12) comes in batts of fibered glass that limit the flow of air into and out of a home. Air becomes trapped between the glass fibers, therefore reducing airflow.

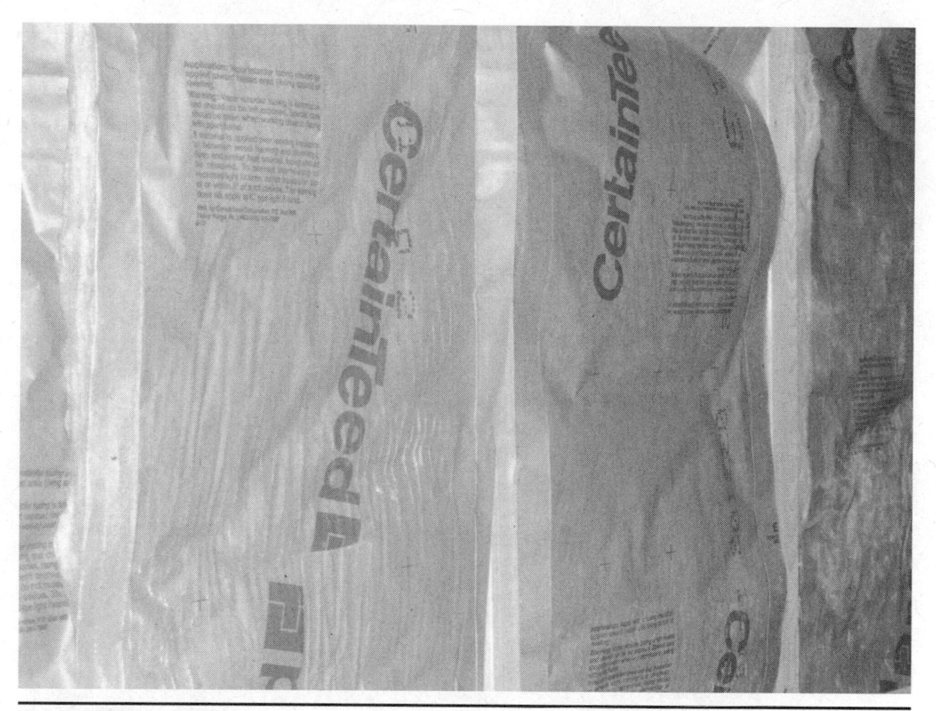

FIGURE 3-11 Insulation.

FIGURE 3-12 Fiberglass insulation.

Rigid-board insulation (Figure 3-13) functions the same as fiberglass insulation but is used in areas where loose-fill or fibrous material cannot be used.

FIGURE 3-13 Rigid-board insulation. *(www.srrb.noaa.gov/surfrad/trailer/figure3.jpg.)*

Loose-fill insulation (Figure 3-14) is made of cellulose, recycled newspaper or clothing, or other materials and functions in the same way as all other forms of insulation, but it can be used in areas where traditional insulation is difficult to install.

FIGURE 3-14 Loose-fill insulation. *(www.cdc.gov/nceh/publications/books/housing/Graphics/chapter_07/Figure7.03.jpg.)*

Spray-foam insulation (Figure 3-15) is an excellent choice for new construction. Spray-foam insulation is very efficiently because it expands and fills all holes, creating an excellent energy envelope.

FIGURE 3-15 Spray-foam insulation. *(www.ornl.gov/sci/roofs+walls/AWT/HotboxTest/WoodFrame/SoyBean/Soy%20Foam%20Wall.jpg.)*

Reflective materials (Figure 3-16), often in combination with traditional insulation, offer a barrier to retard moisture and retain heat.

FIGURE 3-16 Rigid foam board insulation with a reflective coating. *(www.srrb.noaa.gov/surfrad/trailer/figure1.jpg.)*

There is no excuse not to insulate your home. I will identify what types of insulation are correct for you in Chapter 8. In the simplest terms, anyplace where you *can* install insulation, you should.

Install Low-Flow Faucets

Install low-flow faucets and showerheads in your home (Figure 3-17). New low-flow faucets and showerheads can be purchased for a few dollars to as much as a few hundred dollars.

Excessive water usage wastes more than just water. Every time the hot-water faucet is turned on, electricity is used, and heat is produced. The hot-water heater uses energy and fuel to produce hot water. Low-flow faucets and showerheads reduce your water usage by 35 to 50 percent. The savings involves both water and heating costs.

FIGURE 3-17 Low-flow showerhead.

Install Low-Flow Inserts in Faucets

If you cannot afford new faucets and showerheads, low-flow inserts can be purchased for a few dollars and laced into existing faucets.

Plant a Tree

How about planting a tree? Trees can be purchased at low cost or obtained for free. Trees can help to keep your home cool in the summer and warmer in the winter (Figure 3-18). Trees add cooling, moisture, and oxygen to the area immediately around a home. *Real estate fact:* A home with mature trees typically sells for 10 to 20 percent more than one without trees. This is an opportunity to increase the value of your home at little or no cost.

FIGURE 3-18 Oak tree. *(http://photo.itc.nps.gov/storage/images/maca/oaktree.jpg.)*

Install Weather Stripping

Door and window weather stripping is a very cost-effective way to reduce costs and make your home feel warm and cozy (Figure 3-19). Weather stripping can be purchased starting a $3. *Energy envelope* is a new term being used by the building industry. In short, it means that you should seal and insulate all openings to and from your home. This includes cracks, seals, door frames, cable installations, etc.—the entire envelope.

FIGURE 3-19 Weather stripping.

Use Sealant and Caulking Liberally

For anyplace that air can leak from your home that is too small for insulation or weather stripping, use sealant or caulking liberally. Sealant, caulking, and putty all can be purchased for a few dollars (Figure 3-20).

FIGURE 3-20 Weather sealant or caulking.

Use Solar Screens

Solar screens and solar films can provide you with substantial summer savings. Solar screens and solar films help to reflect light and heat in the form of light, away from the windows. The window treatment will reflect a portion of the sunlight, reducing the amount of heat entering the home during the warmer months.

Use Awnings and Screens

Awnings can shade window and door openings, preventing direct sun exposure during warm months (Figure 3-21). Awnings and screens can be purchased in most home-improvement centers. The cost is minimal, but again, the savings are significant.

FIGURE 3-21 Awnings and shades. *(www.roanokeva.gov/WebMgmt/ywbase61b.nsf/85256adf006b014bc1256982003b7e87/74baa6aeb55afc93852573a2005a153c/$FILE/Market%20Awning%20Replacement-3.jpg.)*

Check All Faucets for Leaks

Leaking water is wasteful in many ways. Leaky faucets are a large contributor to water and energy waste. View the total cost of the water system. View the system of water supply to the home. A cold-water leak is less expensive. And a hot-water leak involves wasting water and electricity or fossil fuel. Water is provided to the home under pressure. The water is gathered in a reservoir or pumped from aquifers. The water is cleaned, filtered, and pumped to your home. A cold-water leak does not waste much energy in the home. The water utility, however, will waste the same amount of resources for any leak.

If you have a hot-water leak, then, in addition to your increased water bill, you are wasting some portion of the electricity, oil, or gas needed to heat your water. Your slow drip will cause not only a stain in your sink but also an increase in your utility bill. Many utilities charge more for higher usage. Read your utility bill, and understand how and when utilities are supplied. The act of reading a bill could save you money.

Retrofit all wasteful household faucets with aerators or flow restrictors. These items are very inexpensive, and they reduce the amount of water that is available at the faucet. The performance of the faucet will not be affected, but a significant reduction in water usage will be achieved. These items can be purchased for just a few dollars at any home-improvement or hardware store.

Never install a "water to air" heat pump or air-conditioning system. Air-to-air models are just as efficient and do not waste water. If you currently own one, replace it.

Turn Off Unneeded Utilities When on Vacation

Turn off your water softeners and other utilities while you are on vacation. Water softeners remove minerals that are in water. Commonly, a residence that obtains its water supply from a well may need additional filtration or water softening. Changing the filters at the correct interval also will save energy and allow you to live with clean, drinkable water.

Insulate Your Pipes

Insulate your water pipes. It is easy and inexpensive to insulate your water pipes with preformed foam pipe insulation. One of the benefits will be the availability of instant hot water. This will avoid wasting water and heat while you wait for the water to get hot.

Insulate Your Water Heater

Purchase an insulation blanket for your hot-water heater. This will allow the water heater to maintain a constant temperature and use less energy. Always check with the manufacturer before performing this operation. Newer, energy-efficient hot-water heaters are insulated and do not need an additional blanket.

Replace All Water Fixtures at the End of Their Useful Life

Replace all the fixtures or toilets in your home as needed. Unless you absolutely must have the remote-controlled heated toilet today, replace the units as they become nonfunctional. Table 3-2 demonstrates the amount of water you can save with each new device.

Table 3-2 Water Usage Chart

Conventional Fixture/Appliance	Water Use (gallons)	Water-Saving Fixture/Appliance	Water Use (gallons)	Water Savings (gallons)
Vintage toilet*	4–6 per flush	Low-consumption toilet†	1.6 per flush	2.4–4.4 per flush
Conventional toilet‡	3.5 per flush	Low-consumption toilet†	1.6 per flush	1.9 per flush
Conventional showerhead*	3–10 per minute	Low-flow showerhead	2–2.5 per min	0.5–8 per min
Faucet aerator*	3–6 per min	Flow-regulating aerator	0.5–2.5 per min	0.5–5.5 per min
Top-loading washer	40–55 per load	Front-loading washer	22–25 per load	15–33 per load

*Manufactured before 1978.
†Manufactured from 1978 to 1993.
‡Manufactured since January 1, 1994.
Source: www.mde.maryland.gov.

Do Not Use a Space Heater

If you own a space heater, throw it away (Figure 3-22). These small, inefficient appliances use a great deal of electricity to heat a small area. If you are using a space heater, you probably have not addressed the insulation or heating problems in your home. Spend the money on insulation, not on electricity.

Energy Star and Why You Should

If you are purchasing small appliances, always look for the Energy Star label (Figure 3-23). The Energy Star label does not guarantee that you will save money in every circumstance. Know and understand your needs. In most circumstances, an Energy Star appliance will save significant amounts of energy compared with a comparable non–Energy Star appliance. Be aware of the energy-efficiency rebates or tax credits when buying Energy Star items. A store display explains the tax credits or rebates and usually accompanies the items you purchase. To be sure that you obtain all the benefits from going green, check the Energy Star Web site at www.energystar.gov, or ask your accountant at tax time.

FIGURE 3-22 Space heater.

FIGURE 3-23 Energy Star label.

I am providing you with hundreds of ideas about how to save money and live more comfortably. There are additional worthwhile opportunities that I did or will not address. If you feel that you have an idea that saves energy, send it to me at www.exploresynergy.org, and I will share your ideas.

In Chapter 4, I will address the chief energy expenses in your home. When I say *chief* expenses, I mean the *main usage* by you and your family. Some of the items presented may surprise you. Some items, while used only occasionally, such as a pool pump, use tremendous amounts of electricity but only during certain portions of the year. Comparing your refrigerator to your pool pump, you may find that your refrigerator, while not a significant draw of power, is a continuous drain on your wallet. I will attempt to help you assess the issues that affect your life every day.

CHAPTER 4

Chief Home Energy Expenses

This chapter will focus on the major energy consumers in the home. Identifying and modifying the most costly appliances will save you the most money and energy. Again, it is important to view your home as a system. If the heating system is functioning efficiently but the home is cold during the winter, the solution is not to replace the heating system. The problem is probably a lack of proper insulation.

How Much Electricity Do Appliances Use?

After studying the chart in Figure 4-1, it should be evident why I devoted half of Chapter 2 to the refrigerator. This chart shows how much energy a typical appliance uses per year and its corresponding cost based on national averages. For example, a refrigerator uses almost five times the electricity the average television uses. You can visit www.energysavers.gov for instructions on calculating the electrical use of your appliances.

The first stage in efficiency is to remove any unnecessary appliances. Remove that old second refrigerator in the basement or the deep freezer. Supermarkets today are exemplary at providing fresh, suitably chilled food. Why would anyone want to deep freeze beautiful fresh veal cutlets until they are solid as a rock and as tasteless as cardboard? Are you saving money? Probably not! Why spend extra money to ruin good items? Most people drive past the supermarket every day. Let the supermarkets and the major corporations chill and heat your food for you.

Appliance efficiency doubles approximately every 10 years. When you purchase a new appliance, always look for the Energy Star label. This does not guarantee the most efficient appliance for you, but, on average,

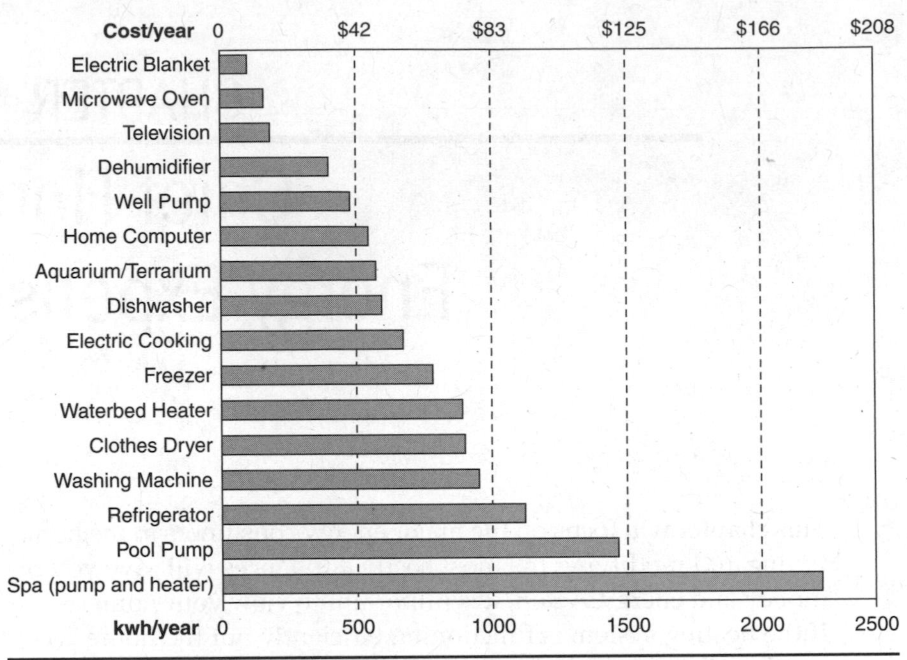

FIGURE 4-1 Appliance electricity-use chart. *(U.S. Department of Energy, January 2009; www1.eere.energy.gov/consumer/tips/appliances.html.)*

Energy Star–qualified appliances will save the average consumer more energy than a comparable new appliance. For example, an Energy Star refrigerator uses a high-efficiency compressor, improved insulation, and more precise temperature and defrost mechanisms to improve energy efficiency. Energy Star refrigerators use approximately 40 percent less energy than conventional models.

When disposing of old appliances, take them to a local recycler. The company will dispose of the item properly and usually pay you a small amount for the appliance.

The Washing Machine

Next on the list in terms of energy use is the washing machine (Figure 4-2). Energy Star clothes washers will save consumers approximately 40 percent over conventional washers. In the figure, can you guess which picture is of an energy-efficient appliance? Actually, they both are energy-efficient, but even I would not use the washboard.

Energy Star–approved washers save substantial amounts of energy and water. Contemporary Energy Star–qualified clothes washers come in

A

B

Figure 4-2 (A) Washer-dryer combination. *(www1.eere.energy.gov/consumer/tips/laundry.html.)* (B) Washboard. *(www.nps.gov/archive/fosc/Graphics/laundress7.jpg.)*

either front- or top-load designs. Both types of units are highly efficient. The most significant design difference for current top-loading washers is that there is no central agitator. Top-loading Energy Star washers use a wash system that flips or spins clothing through the stream of water. Front-loading Energy Star washers tumble clothing through a small amount of water that resides in the bottom of the washer bin. Both designs use less water and therefore less energy. Advances in motor performance allow for more efficient wash and spin cycles and therefore energy savings. Faster spin designs extract more water and allow for less drying time. In addition to the savings they provide, washer-dryer combinations come in many decorative colors and are now part of the room design rather than an eyesore in the basement or closet.

Replace Your Home Clothes Washer

Replace your old washer with an Energy Star model. Today's washer-dryer combination is very efficient and requires very little maintenance. However, as with all appliances, if the units are not maintained properly, efficiency suffers, and costs increase.

Here are some washer tips:

- Only wash full loads. Washing partial loads uses almost the same amount of water and the identical amount of electricity.
- Wash clothing with cold water only. Washing clothing in warm or hot water wrinkles and shrinks clothing. Home heating water is 130°F. This is not hot enough to kill germs; soap kills germs. If the clothing is stained, use pretreaters, or soak the clothing in the water for a few hours and then wash. There is no benefit to washing clothing in hot water.
- Allow the washer to drain at the same height or lower as the washer pump. Do not use the washer pump to force the drainage water above the level of the washer. This is a common occurrence in basements. The washer pump must work harder to remove the water. This will shorten the life of your washer.
- Keep the drain hose free flowing; again, this is to prevent the washer pump from having to work harder.
- Use the regular cycle only. Additional rinses are unnecessary. The additional rinses use many additional gallons of water and extra electricity for no benefit.

The Dryer

Clothes dryers are the next energy-intensive unit on our list. Clothing dryers come in two varieties: gas or electric. Gas dryers are more efficient than electric dryers. The limitation may be that gas is not available in your home. Both types of dryers are energy-intensive.

Replace Your Home Clothes Dryer

The best option when choosing a clothes dryer is to select an Energy Star model that is appropriate to your requirements (see Figure 4-2A).

Here are some tips for an indoor dryer:

- Only dry full loads. This is similar to the washer; partial loads use almost the same amount of energy.
- Clean the lint trap. Allowing the air to flow freely is the key to having clothing dry quickly and efficiently. A blocked lint trap will not allow the humid air to leave the dryer, causing the dryer to run significantly longer.
- Keep the vent and duct free flowing with no sharp turns. Again, this is the same situation as the lint trap. Air must be able to flow freely.
- Clean the ducts. This is to allow the air to flow freely and prevent fires.

Use an Outdoor Dryer

An energy-efficient alternative is to use an outdoor dryer (Figure 4-3). These devices are inexpensive and only cost you money once. Outdoor dryers require zero energy and create zero pollution. They can dry many loads of wash at one time and are easy to use. With an electric or gas dryer lasting about 7 years, a typical person will use 10 dryers in his or her life. With such dryers costing $500 to $1,000, and adding the energy costs, the total cost of ownership may be as high as $10,000 or more. In contrast, an outdoor dryer costs about $50 and lasts a lifetime.

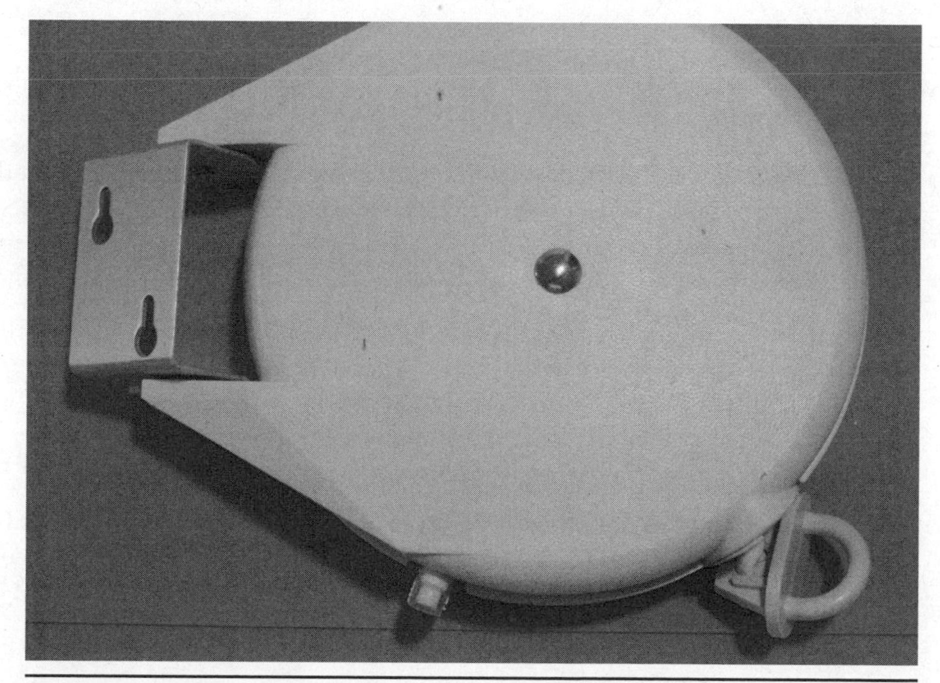

FIGURE 4-3 Outdoor dryer.

These models are sturdy, rust-free, and can dry many loads at one time. You cannot say that for your indoor dryer. Most people can use this option, and for a $50 investment, you could save yourself hundreds of dollars each year. Since the price of oil has risen dramatically, outdoor clothes dryer sales have grown exponentially.

Given all of the above, you do have an alternative to machine-dried clothing (Figure 4-4). Do you remember when you were a child and your laundry was wrinkle-free and fragrant with nature. Today, a retractable clothesline is a transparent, convenient, low-cost option.

Figure 4-4 Clothesline.

Motors, Pumps, and Compressors: Energy Thieves

For the sake of expediency, I will group some of these appliances together:

- Dishwashers
- Room air conditioners
- Dehumidifiers
- Room air filters
- Ceiling fans
- Microwaves
- Toaster overs

These and many other appliances of this type and size are what I call *utilitarian appliances*. These are appliances that do work for you. These appliances are not meant for entertainment but rather to do a job that otherwise would need to be performed by a person.

Purchase Utilitarian Appliances

These appliances all have common attributes. They all use electricity, and their purchase should follow a similar process. First, you should buy appropriately sized appliances. Too large, and you will overheat, overcool, or burn your food and waste energy. Second, always look for the Energy Star label. Energy Star appliances are usually made to higher specifications and use less energy. The energy savings of an Energy Star product usually will pay for the product itself over the lifetime of the appliance. Third, buy to perceived needs or requirements. Understand how the appliance will be used, how often, and for what purpose. Understand why a new appliance is necessary. This selection process will assist you in the purchase of an appropriate appliance.

Hot Water Heaters

Hot water heaters are the most efficient way to create hot water in your home. Today, consumers have the option to choose from many different types of hot water heaters. I will examine the most common and efficient types of hot water heaters available commercially. I will review each type and help you to decide which appliance is right for you.

The residential consumer has five options when purchasing a water heater:

- Solar
- Tankless
- Heat pump
- Gas condensing
- Gas storage (typical or conventional)

Most of the world does not own hot water heaters. Most Americans still have a conventional water heater (Figure 4-5). These hot water heaters are barely more efficient than units sold 20 years ago.

Today there is no need to purchase such a wasteful appliance. The average hot water heater lasts approximately 15 years. If you require a new hot water heater, you should only consider choosing one of the five heater types listed above.

FIGURE 4-5 Conventional hot water heater. *(www.cpsc.gov/cpscpub/prerel/prhtml03/03156a.jpg.)*

Solar Water Heaters

Solar water heaters are the best option. The most advanced solar water heaters require little or no additional energy beyond the sun's rays. The water is heated with sunlight and stored in an insulated container. When hot water is required, it is pumped from the container to the appropriate faucet.

Most consumers are familiar with only one part of a solar hot water system: the solar collectors used to heat the water (Figure 4-8). These are visible roof-mounted units that appear to be thick solar panels. In truth, this is exactly what they are. Instead of capturing the solar heat to produce electricity, the heat is captured and used to produce hot water. Solar hot water systems are an example of design simplicity and efficiency.

Solar hot water systems use the sun's energy to heat water directly or indirectly to heat antifreeze that then heats water through a heat exchanger. A heat exchanger allows the two different liquids to transfer heat but does not allow commingling of the fluids. The heat panels are mounted in any position that will allow direct long-term sun exposure. This type of water heater is known as a passive solar water heater (Figure 4-7).

FIGURE 4-6 Solar water heating panels. *(www1.eere.energy.gov/solar/images/photo_13529.jpg.)*

FIGURE 4-7 Solar water heaters. *(http://pub-lib.jpl.nasa.gov/docushare/dsweb/Get/Version-787/293-8736a_.jpg.)*

An *active solar hot water heater* (Figure 4-8) uses electricity to power pumps to circulate the fluids through the system. Active systems are differentiated in three ways: a direct system that uses pumps to circulate water through the solar collectors, an indirect system that pumps antifreeze through the solar collectors, and a drain-back system that heats the water and allows the water to drain into the holding tank. The active system provides the most utility in mild climates. The indirect and drain-back systems work well in colder climates.

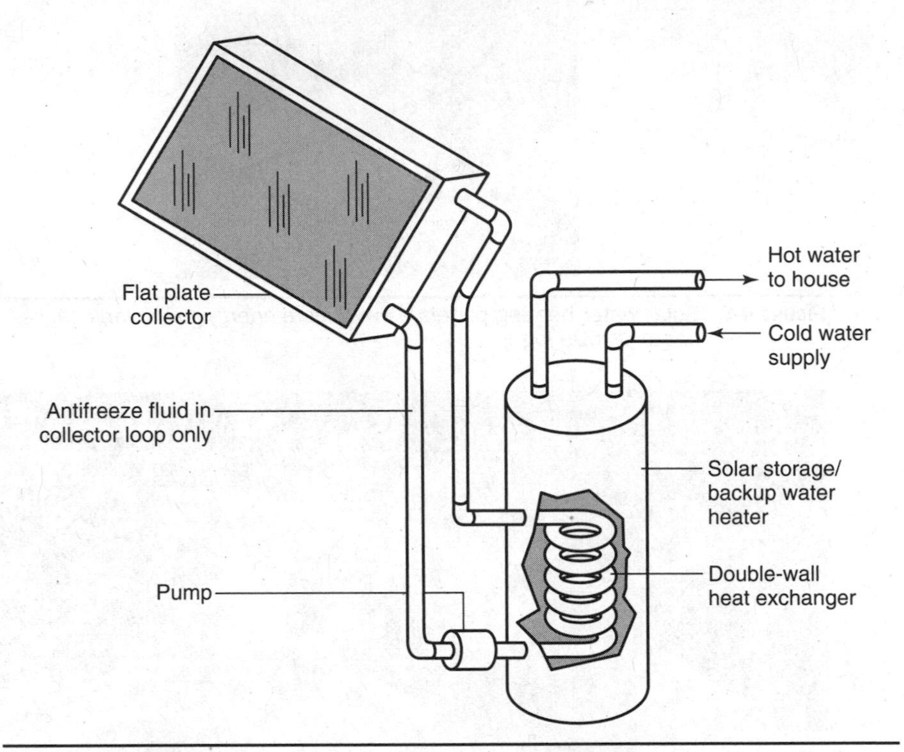

FIGURE 4-8 Active closed-loop solar water heater. *(www.energyeducation.tx.gov/renewables/section_3/topics/solar_water_heaters/img/fig20a_solar_water.gif.)*

This type of product requires knowledgeable and experienced installers. Unless you are familiar with plumbing and confident that you can work at heights, I recommend that you have an experienced professional perform the installation. To help you determine if solar water heat is adequate for your needs, I will cover the basics to provide you with the knowledge you will need to select a system and hire a professional. The requirements for a solar water heater are as follows:

1. Start with your location. Is your building site acceptable for solar heat? Do you have direct sun to the area of installation all year round?
2. Determine the type of system that you need and the capacity that you will require.

3. Have three qualified contractors provide you with estimates and explanations of the potential of solar heaters.
4. Research and locate all tax rebates or credits for your potential system.
5. Ask about the maintenance of each system. Every appliance needs maintenance, and a system such as this is no different.
6. Consider the final cost and whether this type of system fits your lifestyle.

Tankless Water Heaters

Tankless water heaters (Figure 4-9) have become a very popular alternative to conventional water heaters. Tankless water heaters differ from conventional water heaters in that they do not have a reservoir of heated water. Tankless units heat the water on demand. This means that you only use energy when you require hot water. Conventional hot water heaters keep large amounts of hot water ready for your use. This is wasteful because hot water is not being used most of the time.

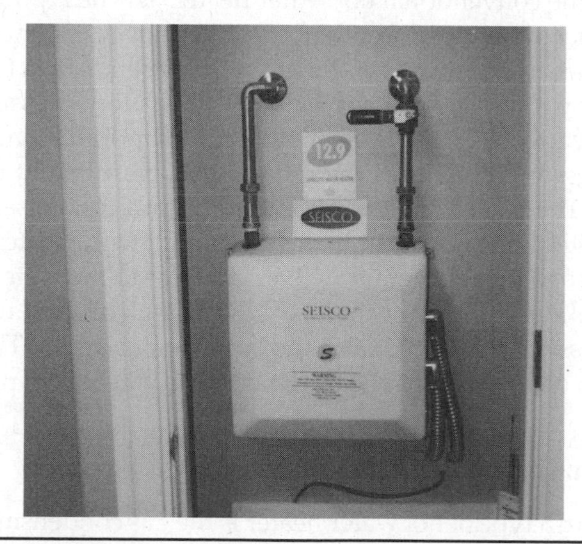

FIGURE 4-9 Tankless water heater. *(http://resourcecenter.pnl.gov/cocoon/morf/ResourceCenter/dbimages/full/592.jpg.)*

Tankless hot water heaters will save you money. These units can provide you with immediate and continuous hot water. No tank is required, so there is also a large space-saving consideration. These units can be powered by gas or electricity and are designed to function for 20 years or more. When you purchase one of these appliances, remember to choose from the selection of Energy Star products.

When installing a tankless hot water system, you have two choices: whole-house or individual unit water heaters. Similar to a conventional hot water heater, a whole-house tankless hot water heater is a single unit that is placed close to the end user. The second option is multiple small tankless units at each location of use. Once you have made this choice, then follow the same rules as for all your other purchases:

1. Educate yourself.
2. Determine what size unit is required.
3. Choose an Energy Star unit.
4. Find a knowledgeable, experienced installer.
5. Check for tax credits or rebates.
6. Review the maintenance required after installation.

Heat Pump Water Heaters

Heat pump hot water systems (Figure 4-10) are another efficient alternative to the conventional hot water heater. The heat pump in a heat pump hot water system, independent of the type of system, will follow the same theory as all heat pumps. A heat pump depends on a temperature differential. Specifically for a hot water system, a low-pressure liquid refrigerant is vaporized in the heat pump's evaporator and passed into the compressor. This causes the refrigerant pressure and temperature to increase. The heated refrigerant goes through a condenser in a storage tank. This will transfer the heat in the refrigerant to the water stored within the unit. When the refrigerant delivers its heat to the water, it cools and condenses. The refrigerant then passes through an expansion valve, where the pressure is reduced, and the cycle begins again. This cycle of warming and cooling is what heats the water in a heat pump hot water system.

Gas Condensing Hot Water Heater

The fourth type of hot water heater is the gas condensing hot water heater (Figure 4-11). This type of hot water heater is very similar to a conventional hot water heater with one exception. A conventional hot water heater vents hot gases as it heats water. The new gas condensing heater does not vent hot gases but rather reuses those gases to add heat energy to the stored hot water. Gas condensing hot water heaters are better and more efficient than conventional hot water heaters. However, the first three choices presented are more efficient options. Choose a gas condensing hot water heater only if there are no other options.

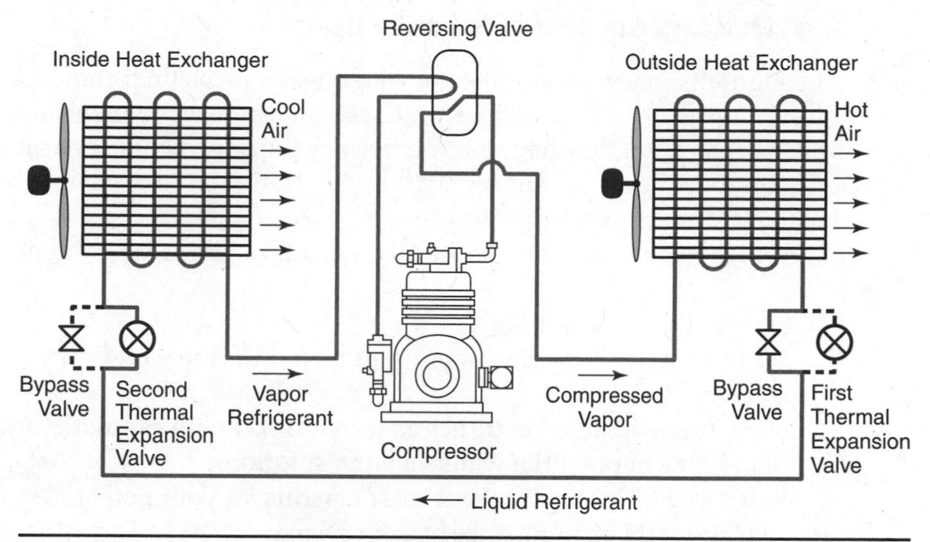

FIGURE 4-10 Heat pump. *(www.cdc.gov/nceh/publications/books/housing/Graphics/chapter_12/Figure12.01.jpg.)*

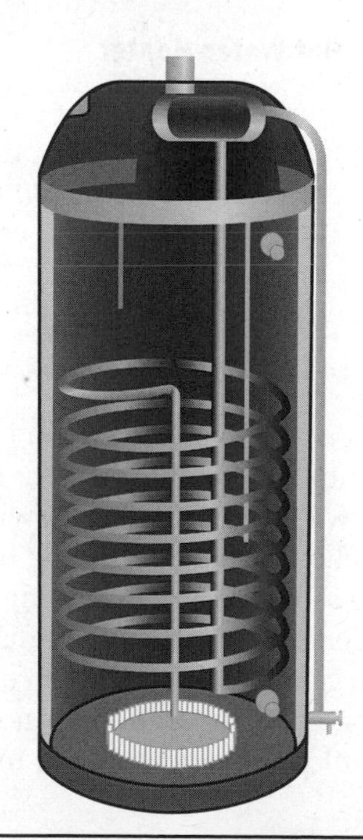

FIGURE 4-11 Gas condensing hot water heater. *(www.energystar.gov/ia/products/water_heat/images/GasCondensing_Works.jpg.)*

High-Efficiency Gas Storage Hot Water Heater

The high-efficiency gas storage hot water heater is the final choice for domestic hot water usage. These units are the same as conventional gas water heaters, but they have a few efficiency improvements. If this is your only option, choose an Energy Star model. In these units, gas is used to heat and store hot water for future use.

Whichever model you choose, always follow the same guidelines:

1. Start with your location.
2. Determine the type of system you need and the capacity you will require.
3. Have three qualified contractors provide you with estimates and explanations of potential water-heating solutions.
4. Research and locate all tax rebates or credits for your potential system.
5. Ask about the maintenance of each system.
6. Consider the final cost and which type of system fits your lifestyle.

Replace Your Home Hot Water Heater

If it becomes necessary to buy a new hot water heater for your home, choose from the first three options above, if possible. These water heaters will provide you with the most efficiency and therefore the greatest cost savings.

Home Cooling

When it comes to cooling your home, you have only three choices: room air conditioners, central air conditioning, or moving to Finland. Okay, Iceland or Greenland will do. If you do not wish to relocate, I will begin to assist you in your choice of home cooling by describing room air conditioners. If you wish to have air conditioning, you should consider central air conditioning. Central air conditioning is more controlled, dispersed, and efficient. If you cannot use central air conditioning because of special considerations or because the property is not owned by you, then a room air conditioner should meet a few simple requirements.

Use a Room Air Conditioner

A room air conditioner (Figure 4-12) is chosen based on a few factors. First, choose a unit that will meet your needs. Your needs are based on the size of the room(s) you wish

to cool and how well that room(s) is/are insulated. If you are renting an attic apartment during the summer, a slightly oversized unit may be necessary for you. Air conditioners are rated in British thermal units (Btu's). Today, most units display the information required on the box—how many Btu's, recommended square feet and/or number of rooms. A rule-of-thumb estimate is that a 5,000-Btu unit will cool one room; an 8,000-Btu unit, three rooms; and a 10,000-Btu unit or larger, one floor of a house. Again, this is a rule of thumb; each situation is different.

FIGURE 4-12 Window air conditioner. *(www.cpsc.gov/cpscpub/prerel/prhtml01/01116.jpg.)*

When buying an air-conditioning unit, notice if the unit requires 110 or 220 volts of electricity. Be sure that you have a dedicated electrical outlet with the correct amount of amperage available. Do not guess about the electrical supply; this could result in a fire. If you rent, ask your superintendent for assistance. When choosing an air-conditioning unit, always buy an Energy Star model. Finally, look for tax credits or rebates before you purchase. A good site, again, is www.energystar.gov. Air conditioning is a luxury, but the government will provide tax credits and rebates so that you can enjoy your luxury efficiently.

Use a Central Air Conditioner

I highly recommend choosing central air conditioning (Figure 4-13) instead of individual room air conditioners. Most energy-sensitive and eco-friendly people recommend natural ventilation, and I agree. However, a few times each year the temperature and the humidly are so

high as to be unbearable. These are the few days of the year that I recommend air conditioning. People with health concerns also require air conditioning. I do not endorse the installation of central air conditioning, followed by an 8-month continuous use of the system. I recommend central air conditioning because in comparison with room air conditioners, there is less misuse, and central air conditioning is much more energy efficient. A properly sized unit and appropriate use dictate the efficiency.

FIGURE 4-13 Central air conditioning.

When purchasing a central air-conditioning unit, the process is the same as for all your other appliances. You should begin to see the pattern here. Installation of a central air-conditioning system is a job that probably will require a professional. After receiving three quotes, you will want to check the contractor and the equipment that he or she is recommending. Confirm the efficiency of the recommended unit at the Energy Star Web site. When purchasing a new Energy Star–rated central air-conditioning system, purchase the appropriately sized unit for your home. Energy Star central air-conditioning units typically will save as much as 30 percent on your energy bills.

If you have an existing central air-conditioning system and it is more than 12 years old, replace it. A new modern system will provide you with better comfort at significantly lower cost. A new Energy Star unit will pay for itself over its lifetime. If you have a central air-conditioning system that is less than 12 years old, have a cooling professional service the unit, and ask him or her how much energy you are using. Then you can decide if it is time to replace your system.

The second part of a central cooling system is the ductwork throughout the home. If you are replacing a central air-conditioning system, I recommend that you install new energy-efficient insulated ducts. If your old ductwork is in excellent condition, have the system cleaned and checked, and then insulate the ductwork. This is also true if the cooling system shares the duct system with your furnace.

Seal Your Duct System

Sealing the duct system is easy but not always successful the first time (Figure 4-14). Anytime the ducts are moved, the entire system can be disturbed. Moving or repairing one length can dislodge a second length of ducting. This can disturb and break fragile seals far from the area being repaired. Leaks can occur in one area after repairs are completed in another area. Confirm that all leakage has been addressed before insulating the ductwork. Insulation will not stop leaks. Air leakage and the pressure differential caused by a leak will only serve to deteriorate the duct insulation and create airborne particles that are hazardous to your health. A cooling professional can validate that your system is leak-free.

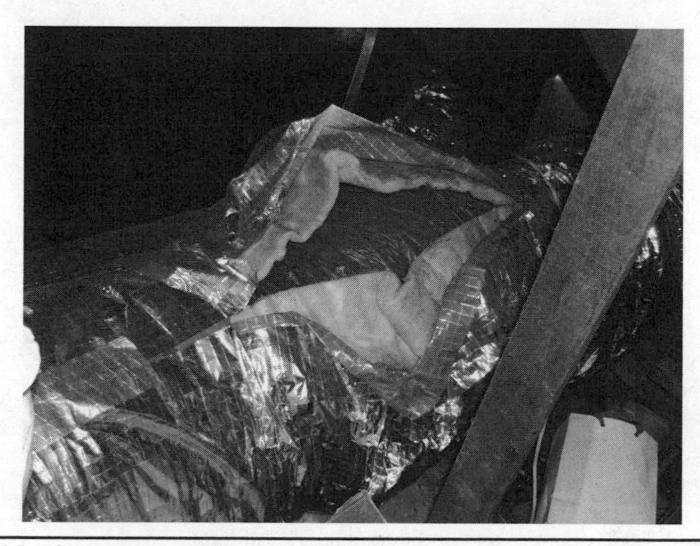

FIGURE 4-14 Central air-conditioning soft duct.

The cooling systems professional can check for leakage using a pressure-differential test. The entire duct system is closed. The specialist installs a high-speed fan to create low or high pressure in the duct system. Colored smoke can be added to the system to assist in the identification of leaks. This type of test is excellent but usually not necessary. System leaks usually are obvious, and most homeowners know when they have a problem.

Perform Regular Maintenance on Your Air-Conditioning System

Now that we have addressed system installation and sealing and insulation of the ducts, the only thing left for you to do is maintenance. Maintenance for heating and cooling systems is very important. You can lose all your efficiency gains if the system is not maintained. Maintenance is simple for most central air-conditioning units. You should have a professional check your system at the beginning of each cooling season. Items to check or maintain include

1. The refrigerant level
2. The filter(s) (Figures 4-15 and 4-16)
3. The ducts (visual inspection) (Figures 4-17 and 4-18)
4. Landscape to keep the condenser dirt- and weed-free

FIGURE 4-15 Central air-conditioning inlet and filter.

FIGURE 4-16 High-efficiency particulate air (HEPA) filter.

FIGURE 4-17 Air-conditioning inlet duct.

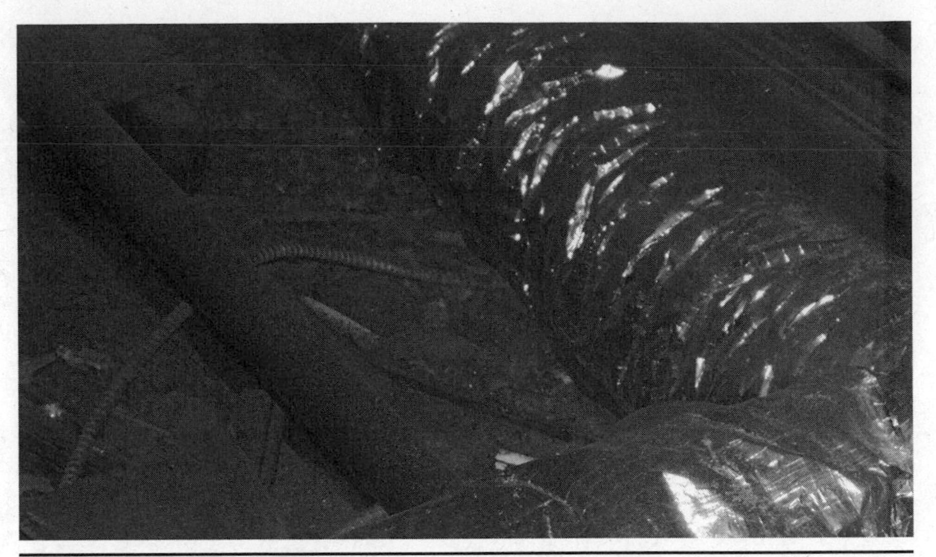

FIGURE 4-18 Soft duct in attic.

Choosing the proper central air-conditioning system for your home and performing quality installation and regular maintenance will allow for years of energy-efficient and trouble-free comfort.

Maintain Appliances

For overall home energy savings, provide regular maintenance for any appliance that requires it. Heating and cooling systems are large, but they are still only appliances, among many others in your home.

Home Heating

The average home will use as much as half its total energy consumption in the form of heating and cooling. When choosing a heating or cooling system or maintaining a current system, the average homeowner needs to be aware of many factors. Consumers incorrectly perceive that the heating or cooling unit is a single entity. The heating or cooling unit is only one part of the total system.

Common residential heating systems come in three forms:

1. A furnace
2. A boiler
3. A heat exchanger

Identify Your Heating System, Understand How It Works, and Solve Problems

What type of system do you own? A furnace heats air and delivers the hot air via air ducts to each room. If you have air vents in each room of your home that provide you with heat, then you have a *furnace*. The furnace is the most common heating system used in the United States. Manufacturers claim efficiencies as high as 97 percent. Energy Star units top the list of the most efficient furnaces.

A *boiler* heats water, and then the heated water is circulated through your home to produce heat. If you have steam heat and radiators or baseboard heating, you have a boiler. Boilers are also very common heating units with high efficiency. Units that meet Energy Star requirements are the most efficient boilers.

A *heat exchanger* is a common form of heating system used in temperate climates. There are two types of heat exchangers: electric air source heat pumps and geothermal heat pumps (Figure 4-19). Electric air source heat pumps are often used in very moderate climates. These units use the difference between outdoor air temperature and indoor air temperature to cool or heat your home.

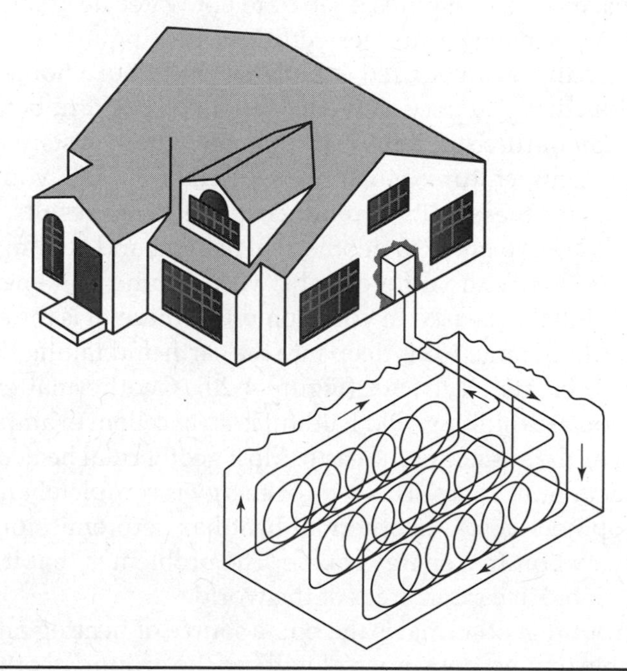

FIGURE 4-19 Geothermal heat pump. *(www1.eere.energy.gov/buildings/residential/images/geothermal_energy_1.gif.)*

Geothermal heat pumps are similar to air heat pumps except these units use the ground instead of outside air to provide heating and air conditioning. These units can and are used commonly to create hot water. Because geothermal units use the earth's natural stable temperature, these units are among the most efficient and comfortable heating and cooling technologies currently available.

Electricity also is an option for heating. Most electrical heating systems are redundant, complementary, or backup systems. Electricity as a heating system is very costly to use. I do not recommend using any type of electrical heating system unless there is no other choice.

What type of heating system is correct for you? This is not an easy answer. Sometimes the heating system is dictated by the building construction or other constraints required. I will assume that you can choose a heating system without restrictions. For a small residential home, small complexes, and small apartment units, a furnace is probably the most efficient choice. With a furnace, optimally the heating unit should be central to the dwelling and have very short duct systems. This will allow for the transfer of heat efficiently to each room of the home or complex.

Boilers are the next efficient choice for heat. If heat must be transferred long distances, for example, in an apartment building, a boiler is the correct choice. Boilers also can be used to supply the hot water for a dwelling. A furnace system requires a separate hot water heater.

Heat exchangers are very efficient heating and cooling systems. A heat exchanger can be used to both heat and cool a home. The immediate benefit is that you need only one system to perform both functions. The limitation of these systems is the climate. These systems rely on a differential in temperature. Much like a Sterling engine, you need a hot and cool area for these units to produce the required results.

Before I explain how geothermal heating and cooling actually works, I need to clarify the difference between geothermal energy and ground geothermal that is used in your home. Geothermal is the energy produced from drilling large wells deep into the earth and taking the hot water and steam to create electricity (Figure 4-20). Geothermal energy is exactly what you are thinking. Old Faithful is an excellent example of geothermal heat near the surface of the earth. Most geothermal heat is deep within the earth near fault zones. This type of energy is completely free except for the development costs. Geothermal heat has zero emissions and is a completely carbon-free energy source. The problem is that it is only economically viable in certain areas of the world.

Ground geothermal is the home source of heating and cooling that is available to most consumers. I will use the ground geothermal unit as an example (Figure 4-21). Ground geothermal energy depends on a temper-

FIGURE 4-20 Geothermal power station. *(www.nrel.gov/data/pix/Jpegs/00427.jpg.)*

ature differential. The inside of your home must be at a significantly different temperature from the ground around your home. Ground temperatures average approximately 60°F. During the summer, you turn on your ground-source heat pump to cool the home. The gas in the heat pump is compressed in the pipes outside the home. The temperature in these pipes will rise to approximately 130°F. Because the ground surrounding these pipes is at 60°F and the pipes are in direct contact with the ground, the heat is removed through conduction. Eventually, the coolant/gas will chill to ground temperature. The gas then is allowed to expand through the pipes inside the home. When a gas expands, it cools, releasing cool energy into the home. The process is repeated until the home is at the temperature requested.

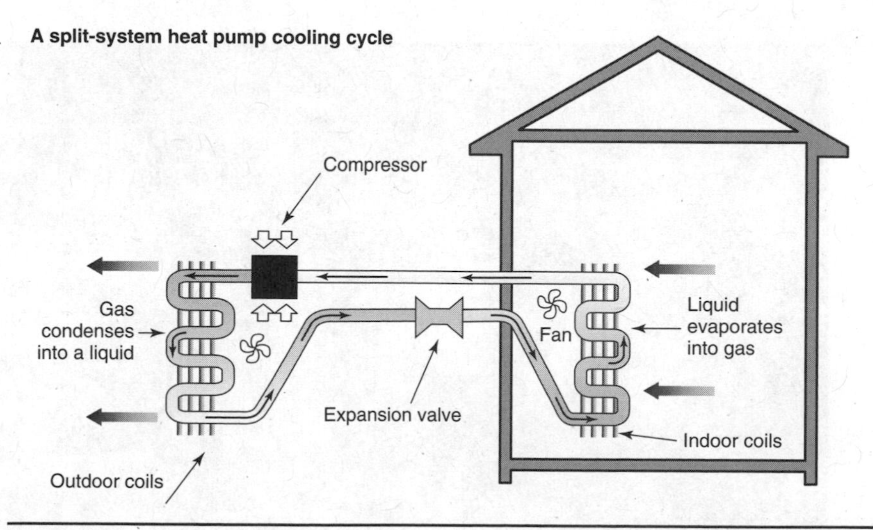

FIGURE 4-21 Heat pump cooling cycle. *(www.energysavers.gov/images/split_system_heat_pump_cooling.gif.)*

In the winter, the system is reversed. The gas in the geothermal system is compressed inside the home. When the gas is compressed, it rises in temperature, heating the home. This system is efficient because it only requires a small compressor to exchange the same amount of gas in reversing directions.

Maintain Your Heating System

Change any filters at regular or required intervals. Poor performance or premature failure of your heating system can occur because of a lack of maintenance. If you cannot do the maintenance by yourself, at least examine the filters and other maintenance items every few months, especially during the months of heavy use. Then call a profession if you observe the need for service.

Yearly maintenance or a tune-up for the heating system is required. You will want to inspect the main unit as well as the inlet and delivery systems. A poorly performing or dirty furnace or boiler can use 20 percent more energy. You can find an excellent checklist for servicing your heating needs at www.energystar.gov/index.cfm?c=heat_cool.pr_maintenance.

Check your pipes or heating ducts. Remember, you want the heat in your home to be transferred directly to the areas in need. If the ductwork leaks, then there will be hot and cold rooms. Sealing and insulating the ducts can increase the energy performance by as much as 25 percent (Figure 4-22).

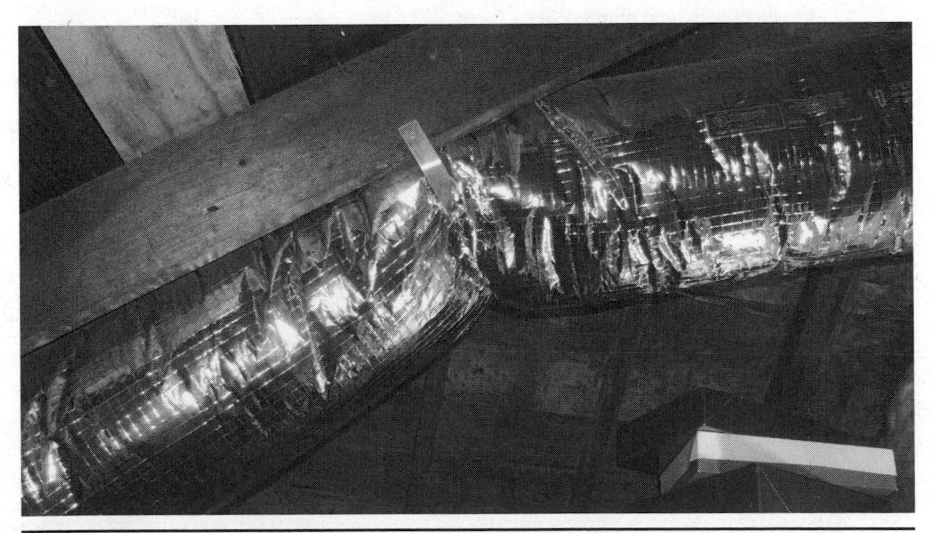

FIGURE 4-22 Insulated ducts.

Seal ductwork to prevent air leakage, and then wrap the ducts in the appropriate insulation. Finally you will be able to use your duct tape for the purpose it was designed. For more information about duct sealing and product recommendations, the Energy Star Web site has an excellent brochure at www.energystar.gov/ia/products/heat_cool/ducts/DuctSealingBrochure04.pdf.

If you are ready to replace your heating system, first choose the type of system that best meets your requirements. Second, find an Energy Star model that will meet your needs. Most Energy Star models are 6 to 15 percent more efficient than conventional units and should save you at least $200 each year. If your system is more than 10 years old, a new system will pay for itself. If your system is less than 10 years old and not providing you with the comfort that you require, have the system serviced by a professional. If the unit is less than 10 years old, it should be efficient and functional.

When purchasing a new heating system, choose a contractor who is qualified to install your system properly. Be sure to obtain the optimal performance from your new equipment. Improperly installed systems could cost you 30 percent or more in efficiency and can shorten the life of your new heating or cooling system (Figure 4-23).

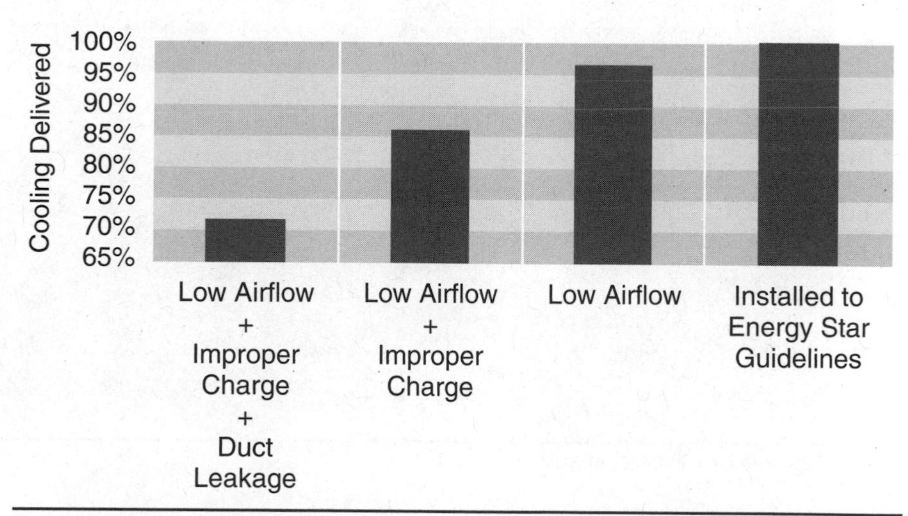

FIGURE 4-23 Quality installation delivers 100 percent cooling; problem installations do not. *(www.energystar.gov/index.cfm?c=heat_cool.pr_hvac.)*

Ask and be aware of what is required for a new heating system. Ask the home installation specialist about the installation. Install the correct size heating unit for your home. If your new heating system is too small, you will be cold; too large, you will be wasting energy. Be sure that the new ducts are sealed and insulated. With a ground geothermal unit, be sure that the coolant is installed, and confirm that the correct coolant is used. Finally, have your installer test the unit and check all the rooms. Look for problems, such as leaks of air, water, or coolant. Confirm that heat is transferred to each room appropriately and in the required amounts.

The final step in the installation of a new heating system should be a programmable thermostat (Figure 4-24). A programmable thermostat will provide you with a comfortable home when you are home and energy savings when you are asleep or away from home.

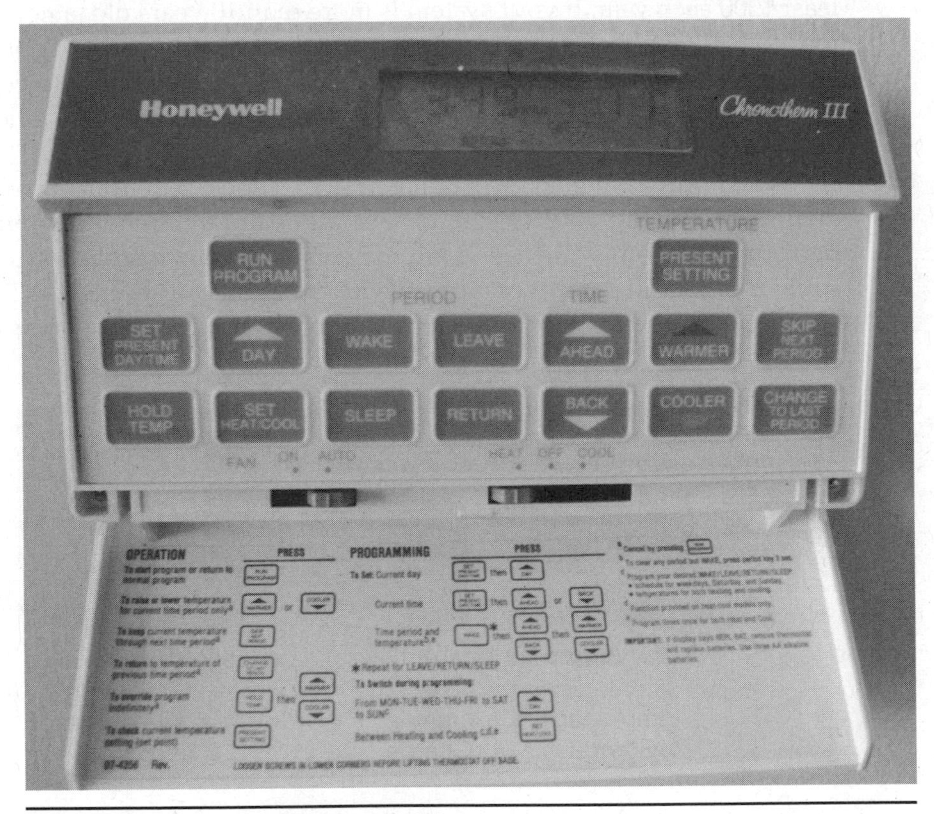

FIGURE 4-24 Programmable thermostat.

Other Large Energy Consumers

The DOE has an excellent checklist for you to use when hiring a contractor to install a new energy-efficient heating or cooling system. You can find this checklist at www.energystar.gov/ia/products/heat_cool/HVAC_QI_bidsheet.pdf.

Reduce Usage of Pool Pump and Other Large Appliances

There are a few large domestic energy-consuming devices that are not typically a part of every household. These items deserve mention because they consume large amounts of power:

1. Pool pumps
2. Aquariums
3. Electric blankets
4. Tanning beds
5. Heat lamps
6. Exercise equipment
7. Fans (do not use them; if it is hot enough for a fan, turn on the AC)

These items consume more energy than the average electric appliance because of the amount of power required or the length of use. Pumps, motors, and compressors all require large amounts of power. A pool pump may require significant power but also may be required to run for 8 to 12 hours. An aquarium may require pumping and heating. What you also should notice is that this is a list of luxury items. These items are not required for basic home use; rather, they are optional for luxury.

Many homes have such items as a treadmill. Exercise equipment contains large, powerful energy-consuming motors. Depending on the habits of the homeowner, it may be more cost-effective to join a gym. Of course, you can always go for a walk; your taxes already pay for the street. The only eco-friendly option that I often see with exercise equipment is that the equipment is used as an indoor clothes dryer.

Now that you understand what uses energy in your home, you can take action to help reduce your utilities and expenses. As always, when buying new appliances, look for the Energy Star label (Figure 4-25).

Figure 4-25 Energy Star symbol. *(www.chandleraz.gov/Content/GB_EnergyStar.jpg.)*

If we review what we have completed, and you have attempted the energy efficiencies revealed in the first part of this book, you should already be experiencing savings and more comfort for little or no cost. These changes automatically translate to global benefits. Next, I will review how you the individual is part of the collective of climate change and how your changes are the greatest changes, both personally and globally.

CHAPTER 5

Personal and Global Environments (They Are One and the Same)

If carbon dioxide (CO_2) were garbage that you could see, touch, and needed to accumulate in your home, you would have been buried in garbage before your first birthday. Humans produce huge amounts of CO_2 without ever knowing or realizing that they have left a carbon footprint. We hear so much about CO_2, but what is CO_2, and why is it bad? CO_2 is a gas at atmospheric pressure that is composed of two oxygen atoms and one carbon atom. This does not sound so harmful. So how do humans produce CO_2, and why is it suddenly bad for the environment?

Global Warming

Let us first gain some perspective on global warming. Since the 1800s, scientists have declared that the release of CO_2 into the atmosphere from the industrialization of nations would raise global temperatures (Figure 5-1). If I recall correctly, this was before Al Gore's birth in 1948. If you do not believe that global warming has been an issue since the 1800s, use Al Gore's other invention, the Internet, to determine the truth for yourself. In 1938, British amateur meteorologist Guy Stewart Callendar explained that humans were responsible for heating up the planet with CO_2. Callendar's main contribution was proposing the theory that linked rising CO_2 concentrations in the atmosphere to global temperature. This eventually became known as the *Callendar effect*.

In the 1950s, the United States made short educational films about CO_2 rise and the effect of temperature rise. The films' producers proposed that global warming would be good. This would allow all frozen areas to become land for agriculture. In the 1960s and 1970s, global

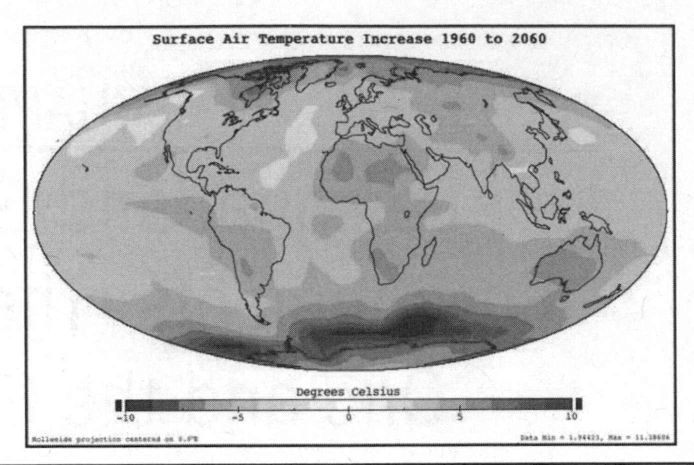

FIGURE 5-1 Global warming. *(www.nasa.gov/images/content/105582main_GlobalWarming_2060_lg.jpg.)*

warming was again a brief topic, and a few hippies did choose to pursue a more productive way of life. Most became yuppies, though, and took corporate or banking jobs. Now, since 2006, Al Gore is responsible for bringing attention to global warming . . . again. This time, however, we must address the issue. We can no longer afford to ignore the problem. We must address the issue not because we are sure of the outcome, but rather because we will cause radical changes to the environment if we don't. Humans depend on a very stable environment for survival. Without a stable environment, hundreds of millions of people will perish. So what can we do personally? Let's look at global warming from an individual perspective.

CO_2 and You

Let's begin with what we *cannot* change. Humans exhale CO_2 and inhale air with oxygen. According to various sources, on average, a human will produce approximately 450 liters by volume or 900 grams by weight of CO_2 per day. How does this compare to . . . anything? One U.S. gallon of gasoline produces 2,421 grams of carbon, and one U.S. gallon of diesel fuel produces 2,778 grams of carbon, according to the U.S. Environmental Protection Agency (EPA).

What else have you done in your life? Have you ever:

- Had a barbeque? The gas or charcoal releases CO_2 in the process of being used.

- Mown your lawn? The gasoline used in the mower, once combusted, releases CO_2. Do not forget the grass clippings; they may be good for your garden, but as the grass degrades, it releases CO_2.
- Trimmed your hedges or cut down your tree? These all release CO_2 as they degrade.

The story of the tree demonstrates perfectly what is occurring on our planet. Normally, a tree will grow for hundreds of years. The tree will consume nutrients from the soil and release water and oxygen as it grows. A mature leafy tree produces as much oxygen in one growing season as 10 people inhale in a year. The tree becomes a *carbon sink*—containing more and more carbon as it grows.

Normally, a tree would reach maturity and die. The tree would begin to decay and return some of its carbon to the atmosphere. The tree may decay and release all its carbon or become buried and slowly decay over thousands of years.

Let us assume that our poor demised tree had a life of 400 years, passed away, remained on the surface, and decayed in only 200 years, releasing all its carbon as CO_2. The total lifetime of the tree's carbon was 600 years. With humans, things are different. Humans plant forests for fast production of wood, paper, firewood, etc. I will use the exaggerated example of firewood.

We have a 50-year-old tree, and we cut it down for firewood, producing CO_2. We ship the logs to a mill using a truck that produces CO_2. We dry the wood in a factory using heat energy that produces CO_2. The wood then is shipped to a store near you, again creating CO_2. Consumers drive to the store, producing CO_2 with their cars. They purchase the wood and drive it home. The wood then is burned, providing the consumer with a few hours of warmth and joy. And as the wood burns, it releases all of its CO_2 at once. As humans, we have taken the natural process of 600 years and reduced that figure to 50 years.

We have accelerated the process of creating CO_2 exponentially. Do you drive a car? In almost half the average American households, CO_2 is produced by the vehicles used.

Do you use heat in your home? Eat, drink, and use water? Okay, I believe that you understand my position. Everything that we do, need, or produce creates CO_2. We cannot avoid producing CO_2 as humans. We can, however, produce less CO_2 and create a sustainable planet. We can sustain our environment by living smarter. What you will come to determine is that if you reduce your CO_2 output or what is called your *carbon footprint*, you will automatically save money. If you do not care at all about the environment or how you will live in the future, reducing your carbon footprint will at least save you money.

Albert Einstein explained that space and time are the same thing. I will assert that money and CO_2 production is inseparable. That is, money equals CO_2. You don't believe me? Let's look at this problem on a macro level. China and the United States are the world's largest economies. Both nations are also the two largest CO_2 polluters in the world (Figure 5-2).

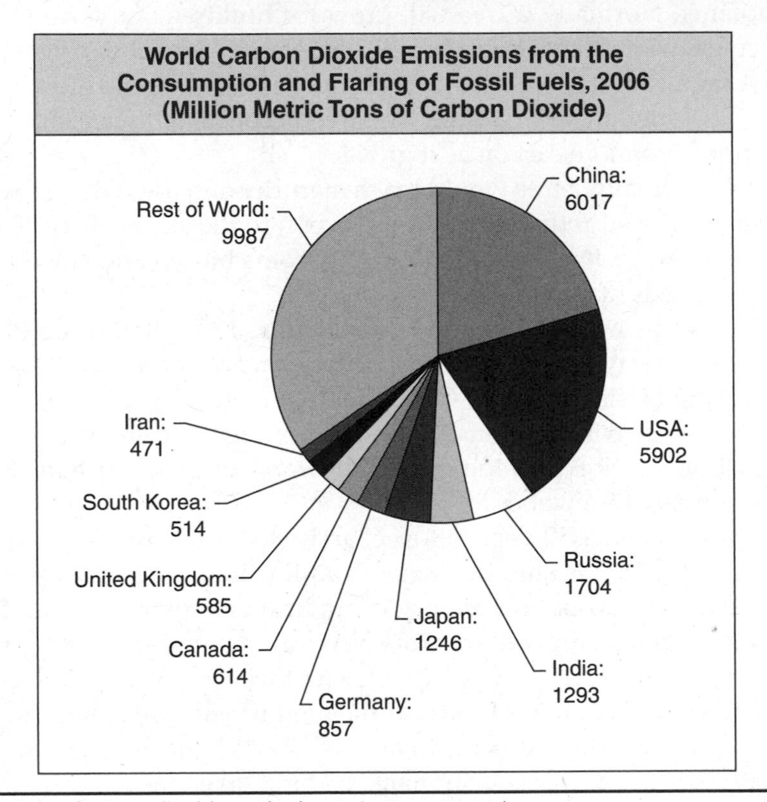

FIGURE 5-2 Carbon dioxide emissions. *(www.epa.gov.)*

If we treat CO_2 like any other pollutant and mandate its regulation, inefficient companies would be forced out of business, and a new business model would develop. I did not think of this idea; this explanation is an oversimplified version of the Kyoto Protocol.

The primary cause of excess CO_2 is overpopulation. If there were only a few hundred million people living on planet Earth, everyone could be wealthy. Everyone could drive the car of their dreams. Everyone could live any way they wished, and their actions would not have an effect on the environment. The natural environment would be able to absorb the CO_2 and all the other pollutants being produced by humans.

The more affluent, educated, and developed people on the earth have low birth rates. There is no need to have 12 children. Placing your resources into one or two children is an intelligent option.

If each couple chose to have two children, the world population would remain stable. Two people are producing two offspring. If two people chose to have one child, then we begin to reduce the world population. Reducing the world population should be the primary world goal.

How can I make the statement that everyone could be wealthy if fewer people occupied the planet? I explained that CO_2 and money are one and the same. An excellent example of this relationship comes to us from the Dark Ages. After the Dark Ages, most modern scientists accepted the fact that approximately 50 percent of the population died from the plague. This left vast resources with no owners. Peasants now became owners of homes and businesses. People of means became wealthy. The very small ruling class and wealthy were the only class that did not completely benefit. The intelligent and wealthy families hired personal armies to take the vacant lands and resources. The less intelligent wealthy were left to fend for themselves. With no skills for this type of life, even many aristocrats were reduced to being farmers. A member of the clergy was noted to say, "Oh how the mighty have fallen."

Without the grimness of mass extinction, we need to reduce the world population, have a more equitable distribution of resources, and care for the earth. Until now, we have abused the planet's resources and left the world to die. Remember, when the earth dies, so do we. Currently, 3 billion people do not have access to clean water or food each day. Population projections show the stabilization of inhabitants of this planet at 9 billion people (Figure 5-3). We know the earth currently cannot support 6 billion people.

Inter-Governmental Panel on Climate Change

The Inter-Governmental Panel on Climate Change (IPCC) was completely wrong with its 2006 projections. The IPCC stated in 2006, as summarized by Al Gore, that sea levels could rise within the next few hundred years. The IPCC projected that the northern glaciers could disappear in a few hundred years. *An Inconvenient Truth* explained that CO_2 levels were rising and that we need to begin to take action. All this information was egregiously incorrect.

Today, science is predicting that we may see all the northern glaciers disappear by 2030 to 2050. The Greenland ice sheet will disappear completely during our lifetime. Not a few hundred years in the future, but

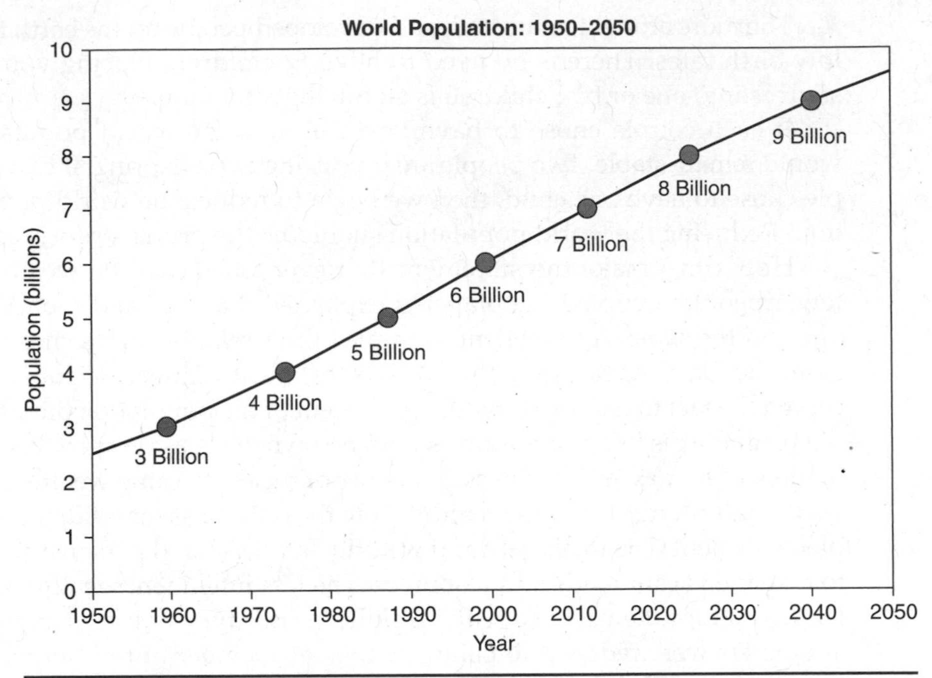

FIGURE 5-3 World population. *(http://images.google.com/imgres?imgurl=http://www.census.gov/ipc/www/img/worldpop.gif&imgrefurl=http://www.census.gov/ipc/www/idb/worldpopgraph.html&usg=__qeiFn92wpASPaE6cNYT7R5n3HzE=&h=481&w=625&sz=11&hl=en&start=1&um=1&tbnid=bAC7E-JLz3D7KM:&tbnh=105&tbnw=136&prev=/images%3Fq%3Dworld%2Bpopulation%2Bprojections%2Bsite:*.gov%26as_st%3Dy%26hl%3Den%26sa%3DG%26um%3D1.)*

now, today. Within the next 21 years, we may see a sea-level rise of 27 inches (68.6 centimeters). More than 330 million people in low-land areas will be searching for new homes. More scientists are beginning to believe that we may be long beyond the "tipping point."

Tipping point is the contemporary phrase that has been adopted to mean the point at which humans will not be able to change the course of global warming and catastrophic environmental damage, including possible human extinction. In fact, I project that we are on a runaway freight train that is still accelerating, with consequences that are dangerous only for the human race. The earth will survive—it always has—humans will not!

I am stating here in writing that we are beyond recovery of a stable planet—that all our efforts, no matter how expeditions or intelligent, cannot prevent the changes that will occur. There is only one way in which we could change our current path: Individuals and individual actions with the assistance of governments and corporations can change our future. Governments alone cannot solve global warming. Corporations alone cannot solve global warming. The only group that can resolve this crisis has its basis in you, the individual human being.

If every individual chose to live green and clean starting tomorrow, we could stop global warming and begin to work toward control of our climate. This may sound arrogant, but if you read this book in 2009, read it again in 2012. I guarantee that all estimates for global warming in 2009 will have been incorrect. In 2012, we will show an accelerated rate of global warming. I require only 3 years to prove this theory.

Humans making the necessary changes will not happen. Why? Since humans began to walk upright, we have had wars. People must choose not to have war. War is highly destructive, and there are no winners. Yet we currently have ongoing wars. If every individual chose not to fight a war, we could not have wars.

How to Change Your Habits

Energy saving in the home is a function of habits. Bad habits cost money; good habits save money. This section will help you to change your habits. It is unfair to ask a person to change and not explain how. The following is a brief and simple path to assist homeowners in changing. If this does not work, place a jar in the most used room of your home. When a person wastes energy, that person must place a quarter in the jar. This is a small accounting measure that will identify how often energy is being wasted.

1. *Pursue change of one habit at a time.* Most individuals at first should attempt to change only one habit. Attempting multiple alterations in one's life may be overwhelming. This will cause the individual to quit attempting change. There is one exception to this rule: If the habits one is trying to change are complementary, then these habits should be modified together. A good example of complementary habits is poor planning and poor eating. An individual's habit of poor planning can cause him or her to buy fast food. By learning to plan, the individual now can purchase and defrost good food for healthy, home-cooked meals.
2. *Create a plan and write it down.* The reason most people do not follow through with changes is that they fail to reinforce and remember to attempt to change. Writing, reviewing, and modifying the plan as necessary will remind you to change. Writing a plan will demonstrate its viability.

3. *Refine or modify your plan.* When attempting any type of change, you are converting from learned behavior to something brand new. Most individuals are not realistic about their goals. Do not be afraid to modify your original plan. As a matter of course, it is mandatory that you revise your plan as necessary as long as you are progressing. Many factors in life will disrupt well-laid pans. What is important is how each us adapts to the changes.
4. *Develop microplans.* Once the ultimate goal is stated and the overall path is defined, develop milestones to allow for positive reinforcement when reached. Microplans can be added, modified, or eliminated as required without changing the entire plan.
5. *Replicate.* Perform the good behavior over and over again. A lifetime of habits will not change without something to replace that undesired habit. Replace the old habit with a new tradition.
6. *Make a commitment.* You must be sure that you are ready to change. If it is not the correct time in life for you to make a change, change the plan or accept that changes will be implemented at the correct time.
7. *Find support.* Do not try to change alone. Employ partners, people who also want to change, or the support of friends and family. The supporters also will act as a reminder.
8. *Write down your motivation and read it every day.* If putting insulation in your home this year will allow you to buy that Tesla Roadster next year, then find a picture of the car, and look at it daily.
9. *Understand the cause of the original behavior.* Avoid situations that will cause the unwanted behavior.
10. *Be consistent with the change.* Do not invent a new way to change every day. Find a system that works, and stay with it.
11. *Be careful of quick success.* Great effort is placed into a new pattern of life. Soon after obtaining success, many individuals slowly regress to former behaviors.
12. *One group of people who are excepted from changing habits are individuals in poor health.* People in poor health often choose goals beyond their temporary abilities. If you are having a bad day, you can forgo your temporary success. I know this from experience.
13. *Sleep is a necessary component of change.* People do not make good decisions when ill or injured. Individuals without the correct amount of sleep are impaired. Maintain your goals by reading your written plan during the impaired days.
14. *Do not quit because of minor setbacks.* Every day cannot be a successful day. Accept that sometimes things cannot follow a strict plan. Do not quit because of a setback. Remember the goal, and try again tomorrow.

15. *The first step to changing habits is to realize that habits are partly unconscious and mechanical.* This is why behavior and thought patterns are not easy to change. Creating new patterns of successful behavior is a long-term but beneficial process. Now close the refrigerator door; you're letting out all the cold air!

The Sky Is . . . Getting Warmer?

If we are "doomed"—why should we even attempt to change?

What else are we going to do? We must do something. What else is more important? The answer is nothing. Nothing is more important than global warming. While the cure for cancer is important in the short term, if there are no people who can contract cancer, why bother? People cannot save the polar bears if humans are extinct. What can we do today as individuals? The first step is to become educated. The second is to take immediate action.

A CO_2 calculator can be found at www.climatecrisis.net/takeaction/carboncalculator/. This and other Web sites will provide standard information. Each person's situation is different, and Web sites such as this one are only guides. The calculator asks for information about automobiles, homes, utility bills, etc. Use calculators such as this one as a guide, not an absolute. Identify how CO_2 is produced in your home and in what quantities. Then create a plan of action to address your largest problem areas. Remember, CO_2 equals money. If CO_2 is being produced, money is being lost.

The carbon footprint of an average person is many times more than just the individual. A 1:1 ratio is required for zero carbon effect. A reduced carbon footprint or less than a 1:1 ratio would allow the planet to begin to recover.

Remember, average human respiratory production of CO_2 is approximately 2.3 pounds (1 kilogram) a day, or 730 pounds a year. Yet globally, the average human produces 4 tons of CO_2 each year, and North Americans generate approximately 20 tons each. Reducing an individual's CO_2 footprint can be as complicated as Figure 5-4. This worldwide greenhouse gas emissions flowchart presents us with a complex picture of how CO_2 is produced and in what quantities.

Today, CO_2 emissions are conveniently packaged with a new name for consumers—the *carbon footprint*. A carbon footprint is the total amount of greenhouse gas (GHG) emissions caused directly and indirectly by an individual (Table 5-1).

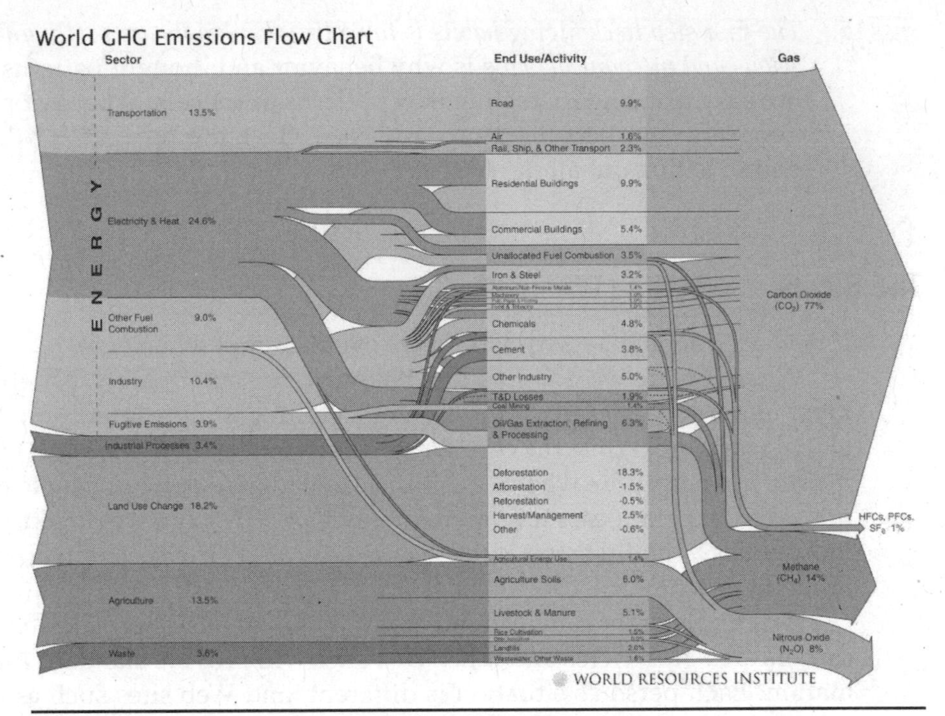

FIGURE 5-4 World emissions chart. *(World Resources Institute.)*

The IPCC was established in 1988 as a sole-source independent provider of information and recommendations about climate change. A complete history and brief description of the IPCC and all documents are free at www.ipcc.ch/about/index.htm. The IPCC does employ scientists to do research but it also compiles information and develops conclusions based on that information. The IPCC is the foremost authority on global warming. All information available at the IPCC Web site is the most contemporary and unbiased information available in the world. Finally, what is the difference between global warming and climate change? In the media, these terms are used interchangeably. In definition, global warming is only one aspect of climate change.

Our Only Hope to Save Ourselves Is You

How can individuals change the planet when governments and corporation cannot? Individuals can change the planet by making better choices. These choices affect governments and corporations. CO_2 is a function of use—or consumerism. The more that an individual does or consumes, the more CO_2 is produced. Typically, it also follows that more money is consumed.

TABLE 5-1 Carbon Footprint

Concentrations and Their Changes			Radiative Forcing	
Species	**2005**	**Change Since 1998**	**2005 (W · m^{-2})**	**1998 (%)**
CO_2	379 ± 0.65 ppm	+13 ppm	1.66	+13
CH_4	1,774 ± 1.8 ppb	+11 ppb	0.48	—
N_2O	319 ± 0.12 ppb	+5 ppb	0.16	+11
CFC-11	251 ± 0.36 ppt	–13	0.063	–5
CFC-12	538 ± 0.18 ppt	+4	0.17	+1
CFC-113	79 ± 0.064 ppt	–4	0.024	–5
HCFC-22	169 ± 1.0 ppt	+38	0.033	+29
HCFC-141b	18 ± 0.068 ppt	+9	0.0025	+93
HCFC-142b	15 ± 0.13 ppt	+6	0.0031	+57
CH_3CCl_3	19 ± 0.47 ppt	–47	0.0011	–72
CCl_4	93 ± 0.17 ppt	–7	0.012	–7
HFC-125	3.7 ± 0.10 ppt	+2.6	0.0009	+234
HFC-134a	35 ± 0.73 ppt	+27	0.0055	+349
HFC-152a	3.9 ± 0.11 ppt	+2.4	0.0004	+151
HFC-23	18 ± 0.12 ppt	+4	0.0033	+29
SF_6	5.6 ± 0.038 ppt	+1.5	0.0029	+36
CF_4 (PFC-14)	74 ± 1.6 ppt	—	0.0034	—
C_2F_6 (PFC-116)	2.9 ± 0.025 ppt	+0.5	0.0008	+22

Source: IPCC Fourth Assessment Report, 2007, p. 141; available at www.ipcc.ch/.

The individual choice of insulating your home could save hundreds of gallons of oil each year. Saving 500 gallons of home heating oil prevents 1.5 tons of CO_2 from being released into the atmosphere. The U.S. Environmental Protection Agency (EPA) has daily homeowner information available at www.epa.gov/otaq/climate/420f05001.htm.

Begin to choose how to live right. Do not allow the government or corporations to choose for you. Corporations are easily circumvented. Corporations are in business to make money. If the corporation is not being responsible, do not use or buy its products. There is international competition for your dollars. Use the power that you have to shape the world. Simple items—calling, writing, and asking the corporation to do the right thing—these measures do work to create change. The reason that corporations do not change is that no one complains.

The government—this is a difficult subject. In the United States, many people are afraid of their government. In Europe, many governments are afraid of their people. And in many parts of the world, some countries do not have a government and some have a government that functions well. Corruption is also an international problem. Corruption, however, follows the same rules as capitalism. Corruption demands the fastest and most efficient way to make money. Corruption can be eliminated or refocused to allow for environmental change that is profitable.

In a viable government, there are only three ways to create change:

1. Support the officials who support a better world.
2. Develop or support candidates to replace bad officials.
3. Vote.

Now we come to the most stubborn, slow-to-change animal in denial—you. More than governments, more than corporations, you, the individual, claim that change is needed. Stop and think, what have you done since 2006 to better your home, your family, and the planet? If you are like most people, the answer is little or nothing.

An individual can decide that the world must change today. An individual can implement change today. In North America and Europe, individuals can choose to use green energy in place of fossil fuels. The purchase of electricity from wind or solar is possible, and the consumer does not have to change any aspect of his or her life. Today, the individual can choose to accept a career in the same town in which he or she lives.

If you, the individual, choose to make these two simple changes today, then the reduction of your carbon footprint could be 70 to 85 percent by the end of the week. So what are you waiting for? You are choosing to heat the earth and forgo any future by not changing. The earth is in need of this type of large change immediately if we humans are to survive. The average person, through the natural process of breathing, produces approximately 2.3 pounds (1 kilogram) of carbon dioxide per day. Everything else you produce is extra.

So far, I have provided you with over 100 completely free changes to better your personal environment. If you do all the free home remedies that I provide in this book *and* insulate your current home, I would wager that you will gain 30 percent in efficiency. This is equal to a 30 percent reduction in utility bills and therefore an increase in your income. The added benefit is that you are helping to save the world, and you are not even a superhero. Add solar power to your home and purchase an electric vehicle, and your average carbon footprint is reduced by 70 to 80 percent. By employing simple measures, the United States could reduce its consumption of fossil fuels by 70 percent. If this is economical, better for the environment, and will cost less money for the homeowner, then why is it not being done?

These options are out of reach for the average consumer. A 10,000-kilowatt solar electrical system could power my home and my new electric car. The difficulty comes with the finances. The purchase of a 10,000-kilowatt solar electrical system will cost $70,000 installed. Over the life of the system, and with rebates, as the buyer, I actually will profit. The problem is that I cannot afford to transition from one type of energy system to the other.

This is where corporations and governments must work with individuals. Governments must provide education, rebates, and consistent tax incentives. Corporations should provide low-interest loans for the installation of green power, products, and service. This will allow for a much expedited transition to clean energy.

Where Do Energy and CO_2 Come From?

In the United States and China, coal is the major source of energy and CO_2 (Figure 5-5). Both countries have large reserves of coal. The coal must be:

1. Discovered
2. Mined
3. Refined
4. Transported
5. Burned
6. Changed to electricity
7. Transported to the consumer by a huge electrical grid
8. Then used by the consumer

Solar energy is delivered free to your home. Solar energy is a much simpler solution with no greenhouse gases. As much as 50 percent of the electricity generated can be lost during transmission to consumers (Figure 5-6).

Oil is the next culprit (Figure 5-7). The oil must be:

1. Discovered
2. Drilled
3. Refined
4. Transported
5. Burned

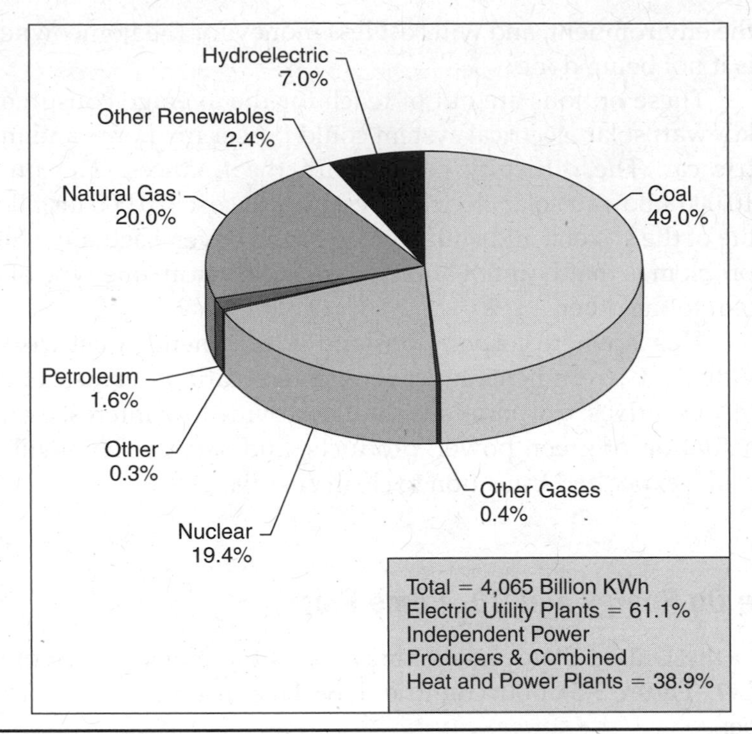

FIGURE 5-5 Power. *(Energy Information Administration, Electric Power Annual 2006, November 2007, Table ES1; available at www.eia.doe.gov/bookshelf/brochures/rep/.)*

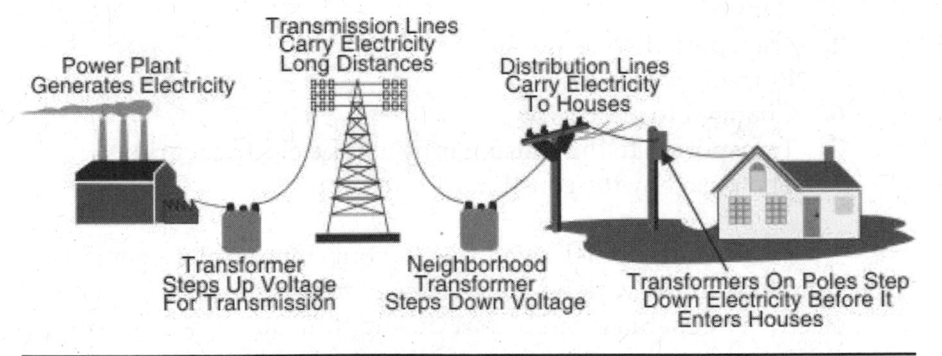

FIGURE 5-6 Electrical transmission. *(Energy Information Administration, www.eia.doe.gov/bookshelf/brochures/rep/.)*

6. Changed to electricity
7. Transported for gas and fuel oil
8. Transported to the consumer by a huge electrical grid or to a gas station
9. Then used by the consumer

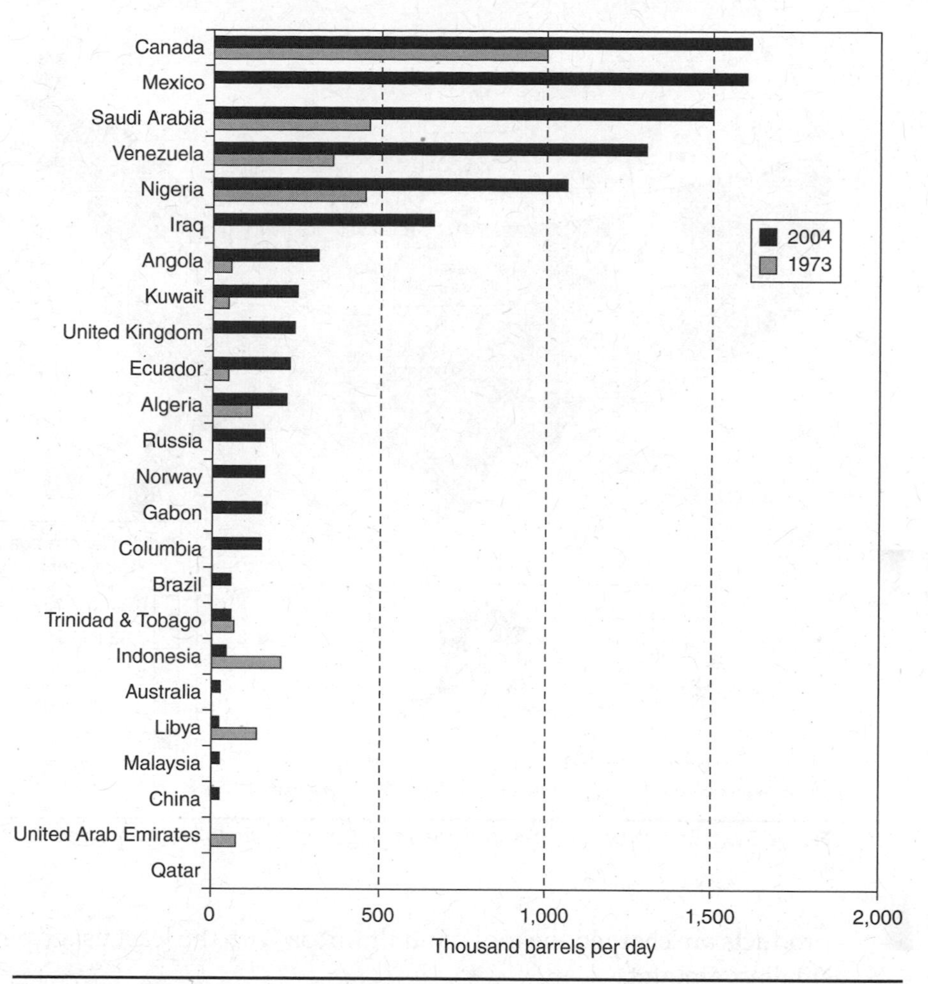

FIGURE 5-7 Oil production. *(U.S. Department of Energy, 2005; www1.eere.energy.gov/vehiclesandfuels/facts/2005/printable_versions/fcvt_fotw375.html.)*

Can this process be more efficient than capturing and using energy directly at a home? The answer is no. Coal and oil are competitive in pricing because the huge infrastructure to support the production of these products is already in place.

Where is energy used? In the United States, Figure 5-8 explains it all.

Electricity is the most needed form of energy. Electrical energy prices are actually in opposition to free-market rules. When a company mass produces an item for sale, to be competitive, the company produces as many units as possible for the lowest price. The company then sells those units to the largest markets for the lowest price. The electric grid and energy companies do the exact opposite. The areas that demand the most

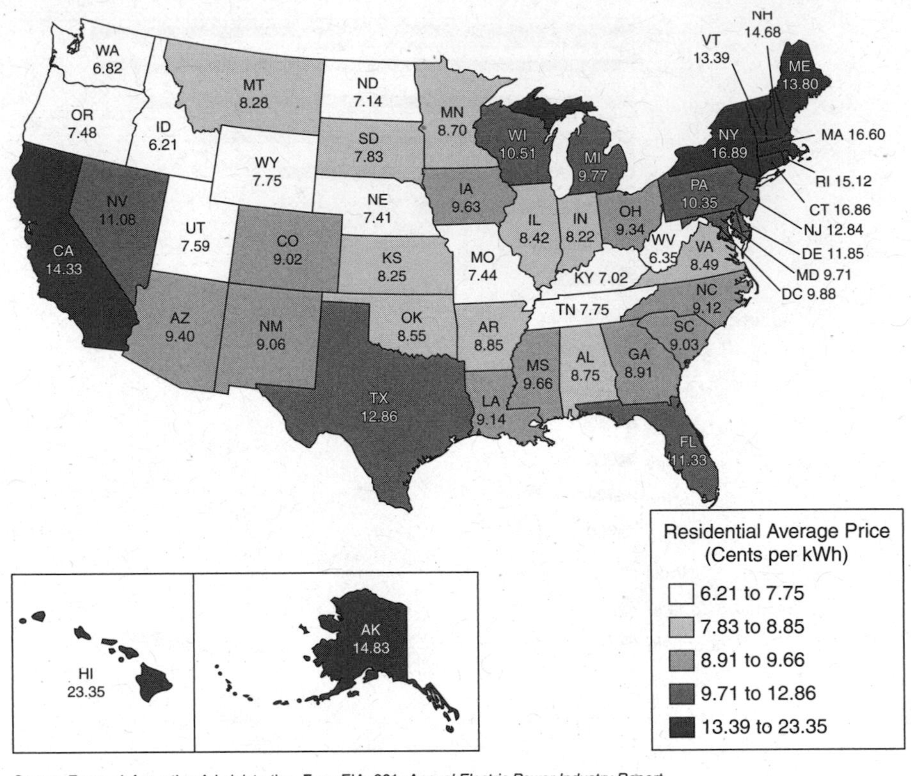

FIGURE 5-8 Energy usage USA. *(www.eia.doe.gov/bookshelf/brochures/rep/.)*

products are charged the most, and the areas with the least usage pay the smallest amount.

An electrical grid is an expensive, unneeded hazard. Wind and solar energy are delivered to your home for free every day.

What is electricity used for (Figure 5-9)?

Generically, the information provided by this figure is good. Now let's move to a more personal or individual question: How are you using your energy?

Homeowners should understand how to read their energy bills. Figure 5-10 is an example of a typical home heating oil bill. The bill reveals the date of the current delivery and the date of the last delivery. This is important because this allows the homeowner an opportunity to see how much oil has been used. The bill shows how many gallons were delivered and at what cost. This bill would have been double in gallons used one year prior. Insulating the attic of the home reduced oil use by 50 percent during the winter months, a savings of almost $450 for each 6-week period that oil is delivered. The savings for this home each year is now thousands of dollars. The insulation cost $100 and created a savings of 1.8 tons of CO_2 that will not be put into the atmosphere each and every year.

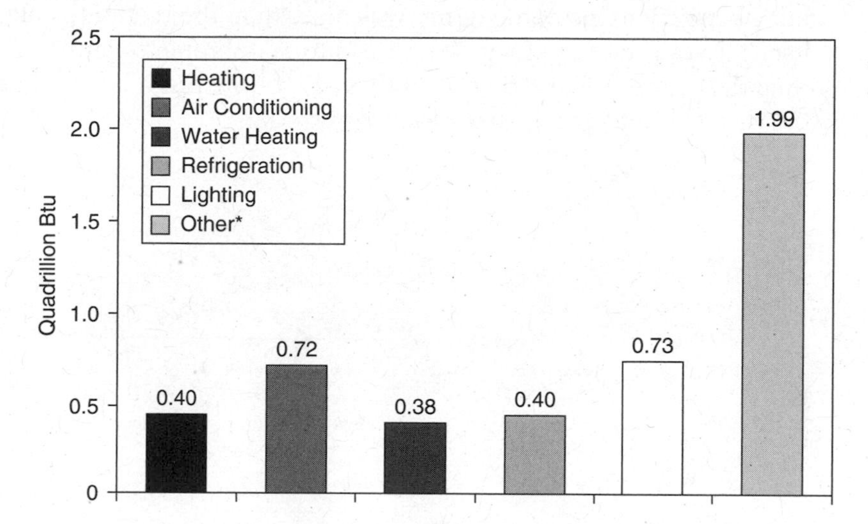

FIGURE 5-9 Energy usage. *(www.eia.doe.gov/bookshelf/brochures/rep/.)*

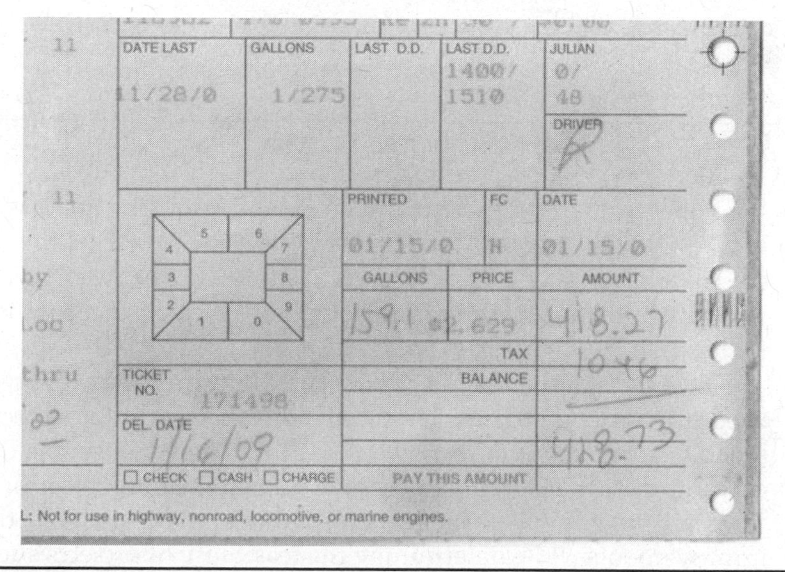

DATE LAST 11/28/0
GALLONS 1/275
LAST D.D.
LAST D.D. 1400/ 1510
JULIAN 0/ 48
DRIVER
PRINTED 01/15/0
FC H
DATE 01/15/0
GALLONS 159.1
PRICE $2.629
AMOUNT 418.27
TAX 10.46
BALANCE
TICKET NO. 171498
DEL. DATE 1/16/09
☐ CHECK ☐ CASH ☐ CHARGE
PAY THIS AMOUNT 428.73
L: Not for use in highway, nonroad, locomotive, or marine engines.

FIGURE 5-10 Typical home heating oil bill.

Electricity is the next largest utility that consumers use. Figure 5-11 shows a typical bill. What is excellent about the display of this bill is that electrical companies are required by law to provide consumers with detailed information. This utility bill demonstrates that the cost of delivering electricity to the homeowner was more than the cost of the electricity itself. If you purchased a pizza for $15 to take home to your family and then the pizzeria charged you another $15 for the pizza box to take home the pizza, would you go back to that pizzeria?

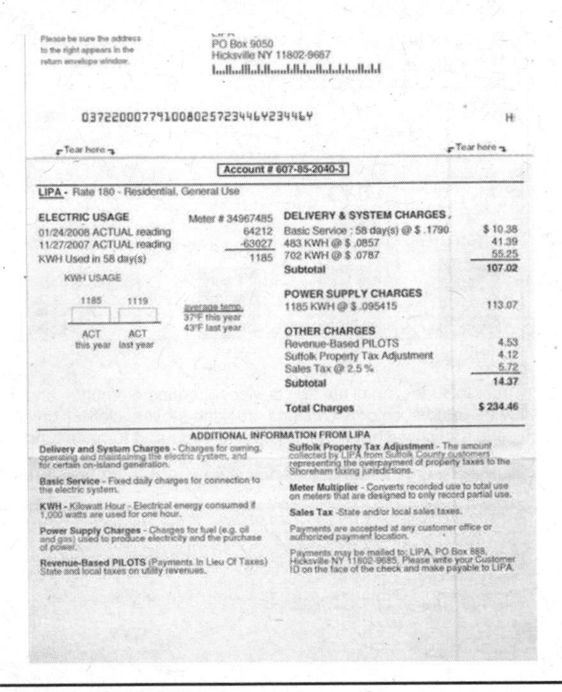
Please be sure the address to the right appears in the return envelope window.

PO Box 9050
Hicksville NY 11802-9687

0372200077910080257234464234464 H

Tear here — Tear here

Account # 607-85-2040-3

LIPA - Rate 180 - Residential, General Use

ELECTRIC USAGE	Meter # 34967485
01/24/2008 ACTUAL reading	64212
11/27/2007 ACTUAL reading	-63027
KWH Used in 58 day(s)	1185

KWH USAGE

1185	1119
ACT this year	ACT last year

average temp.
37°F this year
43°F last year

DELIVERY & SYSTEM CHARGES	
Basic Service : 58 day(s) @ $ .1790	$ 10.38
483 KWH @ $ .0857	41.39
702 KWH @ $ .0787	55.25
Subtotal	107.02
POWER SUPPLY CHARGES	
1185 KWH @ $ .095415	113.07
OTHER CHARGES	
Revenue-Based PILOTS	4.53
Suffolk Property Tax Adjustment	4.12
Sales Tax @ 2.5 %	5.72
Subtotal	14.37
Total Charges	$ 234.46

ADDITIONAL INFORMATION FROM LIPA

Delivery and System Charges - Charges for owning, operating and maintaining the electric system, and for certain on-island generation.

Basic Service - Fixed daily charges for connection to the electric system.

KWH - Kilowatt Hour - Electrical energy consumed if 1,000 watts are used for one hour.

Power Supply Charges - Charges for fuel (e.g. oil and gas) used to produce electricity and the purchase of power.

Revenue-Based PILOTS (Payments In Lieu Of Taxes) State and local taxes on utility revenues.

Suffolk Property Tax Adjustment - The amount collected by LIPA from Suffolk County customers representing the overpayment of property taxes to the Shoreham taxing jurisdictions.

Meter Multiplier - Converts recorded use to total use on meters that are designed to only record partial use.

Sales Tax - State and/or local sales taxes.

Payments are accepted at any customer office or authorized payment location.

Payments may be mailed to: LIPA, PO Box 888, Hicksville NY 11802-9685. Please write your Customer ID on the face of the check and make payable to LIPA.

FIGURE 5-11 Typical electric bill.

Water, water everywhere. Is it any wonder looking at the water bill in Figure 5-12 why people waste water? A total of 21,000 gallons was delivered to this home for $10.50. What is the difference between oil and water? They are both fluids. Water is actually more precious because we can live without oil.

Save the utility bills for your home. Determine how energy is being used in your home. Then determine how energy can be saved. I recommend determining how energy can be saved while improving the lifestyle of the homeowner. Dramatically reducing your carbon footprint does not have to be painful or expensive. I recommend making all free or low-cost improvements while determining the feasibility of projects such as wind or solar energy production.

SOUTH HUNTINGTON W.D.
BOARD OF COMMISSIONERS
Anthony R. Kropp George Kopp Paul Tonna
(631) 427-8190

25 TEED ST

Account Number	Date From	Date To	Meter Reading Previous	Meter Reading Current	Code
23-0001-70	08/06/07	11/05/07	185	206	ES

Consumption 1000 U.S. Gal	Current Amount	Penalty	Prior Bal. Due	Net Due
21	$10.50			$10.50

Next Scheduled Read Date: 02/01/08

TERMS OF PAYMENT Water rentals are liens against the property on which water is used as provided in Subd. d of Sec 198 of the Town Law of the State of New York.

WA = WATER SALES FB = FINAL BILL NO = NEW OWNER ES = ESTIMATED READ
AS OF JANUARY 2001, BILLS ARE NO LONGER GENERATED UNTIL THE ACCOUNT BALANCE IS GREATER THAN $8.00

VISIT US AT OUR NEW WEBSITE: WWW.SHWD.ORG

GALLONS	RATE PER 1,000
0 - 100,000	$.50
100,001 - 200,000	$1.01
200,001 AND OVER	$1.01

FIGURE 5-12 Typical water bill.

CO_2 and the Carbon Cycle*

The final piece that integrates everything in life is another media term—the *carbon cycle* (Figure 5-13). What is the carbon cycle? It is the biogeochemical cycle by which carbon is exchanged among the biosphere, pedosphere, geosphere, hydrosphere, and atmosphere of the earth.

The carbon cycle is usually thought of as four major reservoirs of carbon interconnected by pathways of exchange. These reservoirs are

- The plants
- The terrestrial biosphere (This is usually defined to include freshwater systems and nonliving organic material such as soil carbon.)
- The oceans (This includes dissolved inorganic carbon and living and nonliving marine biota.)
- The sediments (including fossil fuels)

The annual movements of carbon, the carbon exchanges between reservoirs, occur because of various chemical, physical, geologic, and biologic processes. The oceans contain the largest active pool of carbon near the surface of the earth, but the deep ocean part of this pool does not exchange rapidly with the atmosphere.

The carbon cycle demonstrates human influence and how we as a society have unbalanced the carbon equation. The solution is as easy as high school algebra: Balance the equation. The best solution to global warming

*Source: http://en.wikipedia.org/wiki/Carbon_cycle, 2009.

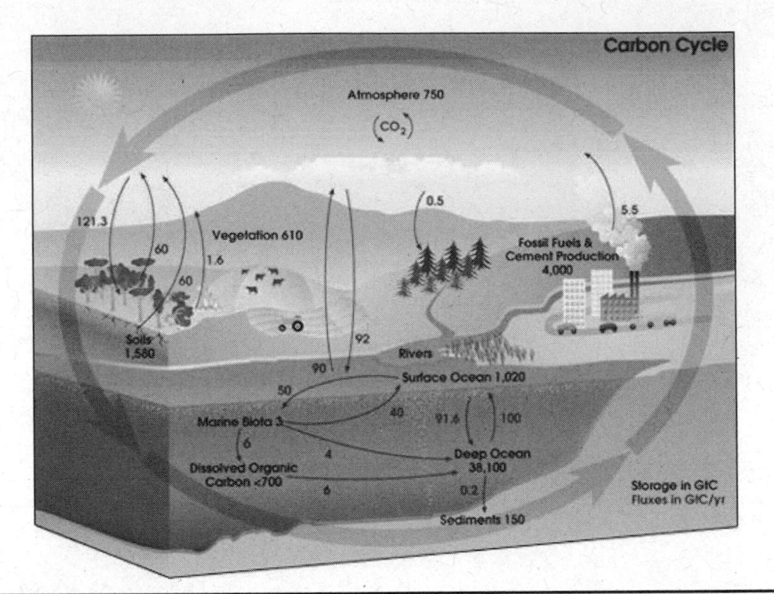

FIGURE 5-13 The carbon cycle. *(www.nasa.gov/centers/langley/images/content/174212main_rn_berrien2.jpg.)*

is for individuals to take action today and help others to begin the process of change.

I will leave you with some less intense information that I am always asked about. The first is why are fuel and oil prices so high?

Federal and state taxes vary somewhat, but for the most part, the oil producers are the clear benefactors of our addiction. The next link is for all of you history buffs: www.eia.doe.gov/emeu/aer/eh/total.html. This link will take you to the history of energy usage in the United States since 1635. This will provide you with a little perspective of the evolution of energy.

To blame any one person for global warming is unfair. To blame everyone is precise. Every individual needs to begin to make the necessary changes to benefit the earth. As we begin to help the earth, a by-product will be that we help ourselves. No governments or international corporations can save the day—just individuals making small changes. The choice is yours.

By improving your life, you are inadvertently improving the life of everyone on the planet. Did you ever want to be a superhero? Today, you can be. Now, how are we going to improve our world? How would you go to China? Would you get into your car and drive? Probably not. You would determine where you wish to go, how to get there, the cost—understand where I am going with this? This first thing that you should do is make a plan.

Home efficiency is no different. Creating an effective plan will save you time, money, and aggravation. This is what I will help you to do next.

CHAPTER 6

Creating a Personal Energy Plan

What does this title mean? What is a *personal energy plan*? How do I create a personal energy plan with my friends or family? Can I create an energy plan for my home? The answer is "Yes." Everyone can create a personal energy plan for their home. Creating a personal energy plan is simpler than most people believe. Start by thinking of your home as a box, with energy into the box and money out. That is,

$$PEP = E - M$$

where PEP = personal energy plan, E = energy, and M = money.

Do you see the relation? The less energy in, the more money the homeowner can keep. This is a very simple formula. Averages can be obtained from your government (Figure 6-1); specifics can be garnered from your own utility bills.

In-Home Energy Use

The key to achieving savings in your home is a total energy efficiency plan. Understand that your home is an energy system with interdependent parts. For example, your heating system is not just a furnace. Heat comes from a delivery system that starts at the furnace and delivers heat throughout the home using a network of ducts or pipes. An energy-efficient furnace will waste a lot of fuel if the ducts are not sealed and insulated properly. Taking a *total home* approach to saving energy ensures that dollars invested are spent wisely.

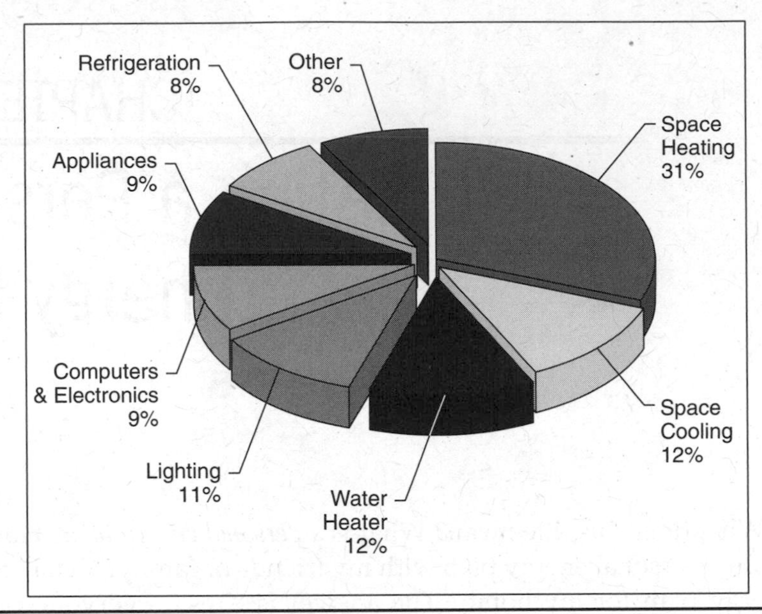

FIGURE 6-1 Home energy use. *(U.S. Department of Energy, 2007; www1.eere.energy.gov/consumer/tips/home_energy.html.)*

Energy-efficient improvements not only make your home more comfortable, but they also can yield long-term financial rewards. Reducing your utility bills will compensate for the higher price of energy. Efficient appliances and improvements will pay for themselves over their lifetimes. In addition, your home will command a higher price when it's time to sell.

Let's take a short break and accomplish some immediate cost savings.

Home Cooling

In the warmer months, do the opposite of what is recommended in the section on heating. The following items will seem familiar because of their direct correlation with the heating section.

Modify Home Temperature Settings

Whenever possible and comfortable, use natural ventilation. When this is not possible, turn the heat down 2 degrees in winter and up 3 degrees in summer. This small change in temperature should not be noticeable. In the colder months, it is often a lack of humidity that will make the home feel cool. In summer, it is the excessive humidity that creates the uncomfortably warm living space.

Understand the problem that is being addressed. If your cooling system turns on often and the home is still warm, the home is not well insulated. Reducing the temperature will only increase the amount of times the cooling system will turn on. The cooler the home becomes, the more often the cooling system will turn on to maintain the temperature requested. If the frequency of the air-conditioning unit is high, then more insulation is required in the home.

Cool air holds less moisture than hot air, which is why it is called *relative humidity*. The amount of moisture in the air is relative to the temperature. This is why people feel cool in air-conditioned spaces. The moisture evaporates from your body rapidly, taking heat with the moisture.

The key here is the temperature differential—the difference between the inside and outside air. Air is always attempting to equalize in temperature and pressure. Your home is a high-pressure capsule that is leaking air to the low-pressure outside world. Insulation slows that process. The higher the differential, the more the air conditioning will be used.

Use a Programmable Thermostat

Programmable thermostats allow you to control the temperature in your home at all times, whether you are at home or not (Figure 6-2). This leads to greater efficiency.

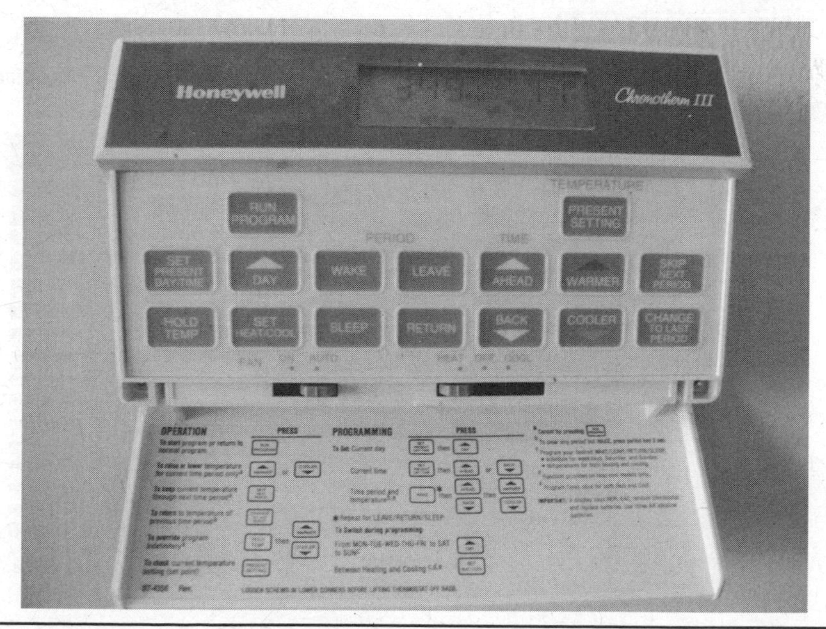

FIGURE 6-2 Programmable thermostat.

Change Temperature Settings When You Are Not at Home

Turn the air conditioning up when you are at work. This is not correct for everyone. It depends on a few variables, such as the efficiency of your home, the temperature, and the length of time you will be away from home. If the home is inefficient and will be vacant for a few hours or more, reduce the air conditioning. If the home is efficient and will be vacant for a long period of time, you will want to make a slight reduction in temperature. Why? If the home is inefficient and leaks a large amount of air, large amounts of energy are being used to cool the rooms when the home is empty. If the home is efficient, the dwelling will not leak much air. Therefore, when the area is vacant, the temperature will not change dramatically. On return of the lodgers, the temperature can be returned to normal without using significant energy.

Clean Heating and Cooling Ducts

Check and clean the insulation and seal the ductwork of the cooling and heating system(s) frequently.

Change Heating and Cooling System Filters

Also check and replace cooling and heating system(s) filters regularly (Figure 6-3).

Figure 6-3 Central air-conditioning system filter.

Heat and Cool Rooms as Required

Close the blinds in morning and open them in the evening, allowing for proper lighting and efficient cooling. Have the air-conditioning system serviced annually. If air conditioning is a requirement, central air conditioning is the correct choice. Central air conditioning has a single unit that provides each room with the proper amount of airflow and therefore cooling. If you have a single wall or window air conditioner, the rooms closest to the air-conditioning unit are cooler than rooms that are farther away. The air condition therefore needs to work harder to distribute the cool air around the home. The placement of these types of units is important. Until you add central air conditioning to your home, place the larger individual units in a central area where they can cool most of the home.

Keep Ducts and Vents Clear

Inspect air ducts and vents, as well as insulation, for leakage annually. Keep the condenser clear and clean. Cool only the rooms in daily use; close vents and the doors of the unused rooms to keep cool air where it is needed the most. Keep air vents clear to allow air movement into each room.

Ventilate the Attic Properly

Ventilate the attic to keep the house cool during the summer months.

Place Air-Conditioning Units in the Shade

Position any air-conditioning units out of direct sunlight, preferably in the shade.

Seal Home Air Appliances

Individual air conditioners should be removed at the end of the warm months, and the breach should be insulated. Caulk any holes that enter the home around the unit. If you use a heat pump, have a certified technician service the unit annually.

Strategic Energy Plan

How to Create an Effective Energy Plan

Creating an effective efficient plan is the most important part of your energy-efficient improvements. A viable plan will save time, money, and future confusion and problems. Address what costs you the most; this will save you the most money and therefore will also save you the most energy. An added benefit is that the money saved can be use to implement the next improvement.

Create a strategy for the largest energy consumers found in the home. Determine the most efficient solutions, and then begin addressing problems. Overall, this is the plan. A proper plan is required to be very detailed. I would not encourage you to undertake such a large project as planning without some assistance.

The information provided will allow for a wide range of opportunities to reduce energy consumption and its impact on the environment (Figure 6-4). The amount of information presented can be overwhelming. Planning is the best way to begin to accomplish the goal. Begin the journey in reverse. If the goal is energy improvement, home improvement, or an increase in income, then the plan will be simple. All these improvements are the same. If the starting point is elusive, begin with some experts. Begin the project by interviewing contractors. The expertise to evaluate your home and create a plan will be revealed by the home-improvement specialist. The plan will include the successful implementation of goals. This is true even if a remodeler will not be used.

Taking action is the first step. Begin with your goals; use a professional to assist with your plan if necessary. Then begin to develop the plan. What follows are the steps for creating a successful strategic plan. A strategic plan is the same whether you are developing a green renovation or planning a trip to the moon. The steps or procedure for creating the plan do not vary; only the details change.

It is often difficult to begin a new project. A clear set of goals, activities, and milestones will assist you to begin and stay on target. Goals are important because you need to know what is to be accomplished. Activities are important because they help to establish the process for achieving the goals. Milestones present you with accomplishments along your path. Realizing that progress is being accomplished allows for a sense of satisfaction and relief.

Determining the clear, well-defined goal is one of the most important steps. This will allow you to move your focus to how the goal will be ac-

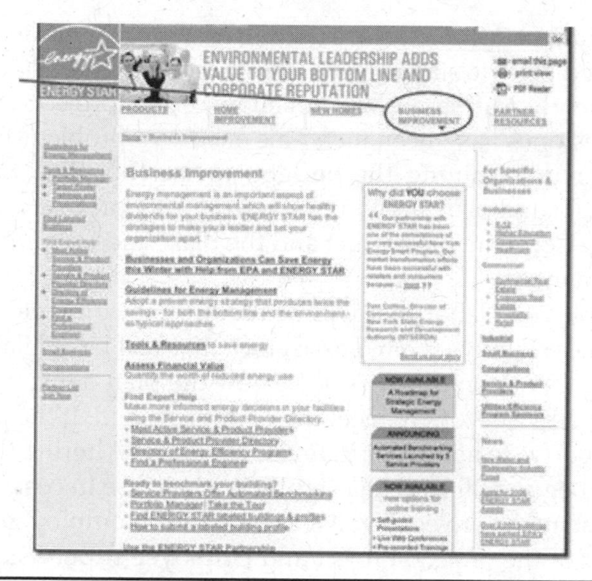

FIGURE 6-4 Personal energy improvements. *(www.energystar.gov/ia/business/healthcare/images/screen2.jpg.)*

complished. This process then will reveal what efforts will be required to accomplish each task. Let's begin.

- Write down a clear, well-defined goal. If this goal or change involves your family, compose the plan with their assistance.
- Determine the budget, and be accurate. Most remodels cost more than anticipated. Budget 10 percent more than determined, and do not change your budget to meet your plan. Change your plan to meet your budget.
- Determine what is required to achieve the goal. Assemble the necessary information about the project and activities. Determine what actions need to be accomplished. Determine the order in which all activities must proceed. Do not put in the new carpet before the new paint. A discussion with an expert may be useful at this time. Having expert assistance may prove invaluable. Obtain cost estimates on the project to determine the budget. This is the time to put the project on paper.
- Establish an actual timeline for each activity. Any contractors will have supplied this information. If you are using a remodeler or contractor, use the average completion estimates, but know the worst-case scenarios. Plan on the average timeline, but be prepared to live with the worst-case scenario.
- Establish milestones or targets of work completed and dates. Milestones demonstrate the progression and allow for a feeling of accom-

plishment. Milestones are used to ensure that certain activities are completed prior to starting new ones. Some activities can or will occur concurrently. Milestones will not allow certain projects to begin until other milestones have been met. This ensures a problem-free plan.
- After determining the budget, revise the plan. Synchronizing the budget and plan at this juncture will resolve large difficulties in the future. Compare the budget and the plan, and be sure that they are complementary.
- Employ the labor that will be required to complete the goal. Confirm friends, family, or hired assistance that will be available for the project.
- Begin work on the project. Review the plan daily or as often as necessary. Revise work to meet the plan as needed. If the plan was designed properly, do not revise it; revise the work. Altering the plan once work has begun will lead to a significant increase in cost.
- Manage the project. Even if a contractor is employed, you are the manager of the project. Project and employee supervision will be a necessary daily task.
- Manage the budget. The budget should include milestones and match the progression of the project.
- When the project is completed successfully, the work is not done. Review the budget, and be sure to collect refunds, rebates, and tax benefits applicable to your project.

Creating a plan is a skill that anyone can use. Apply this information to more than a home energy audit and see how effective planning can be (Figure 6-5).

FIGURE 6-5 Help. *(www.publicdomainpictures.com.)*

Personal Energy Plan

The skills provided to develop a general plan now can be used to create a home personal energy plan. Developing an energy plan is simple. The goal is to save money by reducing your utility bills. How to accomplish this goal depends on the needs of the home and your personal requirements. The homeowner can insulate the attic or choose a slightly modified version of the same plan, for example, to vault all ceilings in the home and insulate the underside of the roof. Both renovations accomplish the same goal, but the latter is more expensive and done for aesthetic as well as efficiency purposes.

Before making any plans, be in a good mood. This may sound crazy, but it is true. Do not make decisions while you are feeling ill. If you are sick, injured, angry, or subject to any other emotional or physical trauma, do not make the final decision until you are feeling well. Planning can occur during a difficult time in one's life. When the individual is feeling better, then the plans can be reviewed and the final decisions determined. This is especially true for large or expensive projects. If possible, make the decisions with trusted friends or family.

The first step in a personal energy plan is to identify the most significant expenses in your home. The largest expenses clearly will demonstrate where most energy is being lost. During energy audits, the foremost problem identified is heating and cooling expenses. These expenses are almost always very high owing to insufficient insulation in the attic. Rarely, the heating or cooling system is the problem.

This particular "problem" is a good problem. It is easy to remedy, and with store rebates, tax rebates, and tax credits, the insulation may cost you nothing. Doing the installation yourself will reduce cost further. Hiring a professional will cost less than $1,000 and usually will require less than a day. The best part about this remedy is that the homeowner will save hundreds or thousands of dollars each and every year.

I will assume that your energy expenses are unknown. The first step is to gather all the utility bills. Write down the total of each bill for the year. Determine:

1. How much money is spent on each energy bill?
2. Which bill is the largest?
3. Will the remodel be done by a contractor or by yourself?
4. How much time will the remodel take?
5. How much will the resolution cost?
6. How long will it take to recover the money spent to repair the problem?

7. How soon will the repair begin to increase your income?
8. How many years until the remodel increases your income?
9. How long do you plan to own your current home?
10. What future home-efficiency projects do you want to undertake?

This is a good beginning for your personal energy plan. Efficiency remodels are complementary. Insulating the attic is excellent. Insulating the attic and installing energy-efficient doors and windows is significantly better.

Creating an entire home plan is even more efficient. This will allow for aesthetic improvements and proper plan development. Install insulation in the walls at the same time that windows and doors are being installed. This may seem obvious, but many homeowners will choose to do one remodel each year. Having a clear plan of the total home will save time, money, and difficulties.

Prioritize your energy resolutions with the whole house efficiency plan. The plan will allow for a strategy when contracting and making smart purchases. If windows are the next item on the home energy plan to be completed next year, and a contractor or store is having a sale on windows this year, then the purchase can save additional money from the already delegated budget. If the energy plan indicates that the windows will not be completed for another 3 years, the homeowner can forgo the sale. Technology will continue to create better and more cost-effective products, and storage of windows for many years does not make sense.

If you are doing the entire home remodeling yourself, it is still a good idea to get the advice of a professional. Many utilities conduct energy audits for free or for a small fee. A professional contractor will analyze how well your home's energy systems work together and compare the analysis against your utility bills. This will support your personal home audit results. The contractor has tools available that he or she will use. For example, equipment such as blower doors, infrared cameras, and surface thermometers will find leaks and drafts. The contractor or auditor will give you a list of recommendations for cost-effective energy improvements. A reputable contractor also can calculate the return on your investment in high-efficiency equipment compared with standard equipment. In some situations, a contractor performing the improvement actually may save the homeowner time and money.

The planning is complete, the budget is known, and the remodeler is you. The next step is to take action. Purchase the amount of insulation required for the project. Actually, always purchase 10 percent more materials than required. Purchasing extra materials will ensure that the project will be completed without interruption. The additional material, if not required, can be returned for a full refund. Then perform the installation.

When the work is complete, the job is not done. Confirm that the installation has been a success by testing. This can be accomplished with direct testing or comparison with normal use over time. Finally, apply for the tax credits and rebates that you deserve for saving the planet.

The insulation of the attic is complete. The home is 30 percent more efficient, and the rebates have been recovered. Now is the time, while the strategic planning and implementation skills are available, to plan the next energy project for your home. The second home project will benefit from the lessons learned in completion of the first. This is also the time to begin to look at large alternatives for the home, such as wind or solar energy.

The home energy improvement process is a self-fulfilling prophecy. The more improvements you complete, the more money you will save. The more money you save, the more projects you can complete. Eventually, there will be no more projects, but there will be additional money.

Items to Include in Your Energy Plan

Home electronics and home office equipment should be outlined in a home energy plan (Figure 6-6). These items individually do not consume a significant amount of power. Home entertainment items do have four very bad habits, though:

1. Homeowners usually do not purchase these items with energy savings in mind.
2. Modern homes contain hundreds of these items.
3. Most of the home entertainment appliances consume vampire power.
4. All of these items are luxuries.

Because of these four issues, home entertainment appliances deserve special attention. This is a short list of these products in your home. Do not forget to multiply them to account for all the people in your home.

1. Battery chargers
2. Cordless phones
3. DVD players
4. Cable box/digital converter box
5. Audio, radio, stereo, iPod, CD players, receivers, speakers
6. Home theater systems and surround sound
7. Power adapter transformers to convert 120 volts ac to 12 volts dc

8. Televisions (if you have plasma types, etc., time to upgrade to digital TV)
9. Computers
10. Copiers and fax machines
11. Digital duplicators
12. Notebook computers/tablet PCs
13. Monitors/displays
14. Printers, scanners, and all-in-ones
15. Water coolers
16. Routers
17. Scanners
18. Cell phones
19. Digital cameras
20. Video cameras

FIGURE 6-6 Home energy plan.

Still doubt that you have hundreds of these items in your home? All these items require power and (some) vampire power. The only successful strategy is planned purchase and distributed usage. The best example of this is the DVD player. Most families have one DVD player in each room for each individual. A smart purchase would be a DVD player that has multiroom access.

Plan your purchase for shared use. The DVD, home entertainment, movie, music, and home office products can be centralized so that all family members can use these luxury appliances but not interfere with the usage of any other family member. Music and movies are now digital. Place all your DVDs and CDs onto your computer. All family members will be able to access them any time they want. The space savings will be significant. The energy savings will be greater. By doing this, the average homeowner will be able to remove 10 or more redundant devices from the home.

When purchasing new products, always look for the Energy Star label.

Place All Chargers in Surge Protectors

Solution #119

Home entertainment and adapters and battery chargers can be plugged into power strips with surge protectors (Figure 6-7). When not in use, all items should be turned off completely. Power adapters convert 120 volts to 12 volts. These items do this continually. Charge your cell phone, laptop, iPod, and other portable items in the car. The electricity is free, almost.

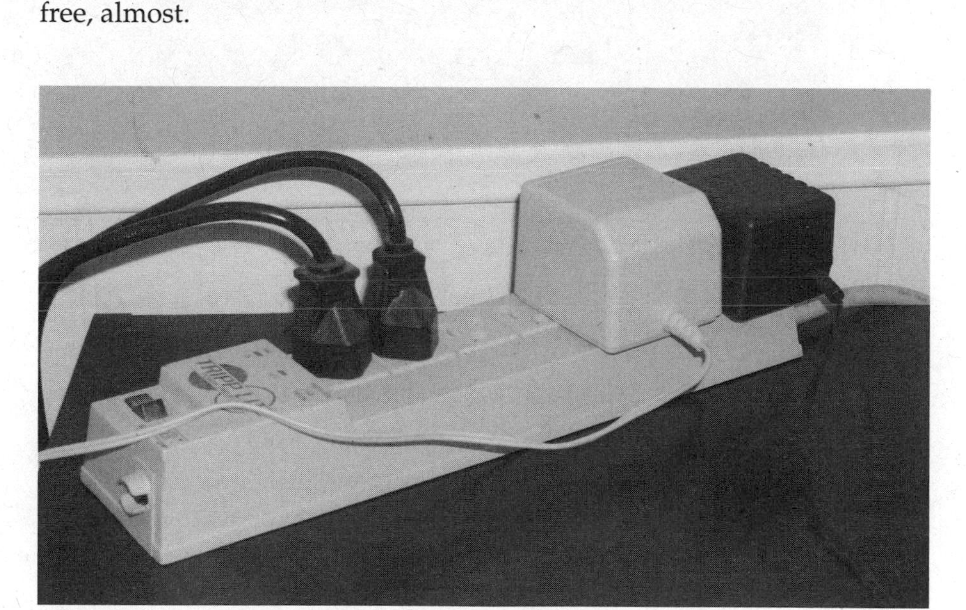

FIGURE 6-7 Surge protector.

Buy All-in-One Devices

When purchasing office equipment, attempt to purchase all-in-one devices (Figure 6-8). Ask your cable or service provider for energy-efficient set-top boxes. The television is no longer a single device. Purchase a cable- and Internet-ready monitor for all your needs. Why have a TV and a large computer screen in the same room?

FIGURE 6-8 All-in-one device.

The following is an interesting note from energystar.gov.

Plasma, LCD, and Flat-Screen Displays

Both plasma and LCD [liquid-crystal display] use different picture-generating technologies than the standard CRT, making them lightweight and super-thin in comparison. However, the term "flat screen" can be confusing because CRTs can also have flat screens.

Plasma TVs are known as "emissive" displays because the panel is actually self-lighting. Basically, the gas (plasma) causes the pixels to glow, which creates the TV image.

LCDs are known as "transmissive" displays because the light isn't created by the liquid crystals themselves. Instead, a light source (bulb) behind the panel shines light through the display.

Flat Screen Display strictly refers to the flat surface of the TV screen. It does not necessarily refer to plasma TVs or thin, flat-paneled LCDs. Flat-screen CRT sets are available and are usually less expensive than other flat screens such as plasma and LCDs but tend to be more expensive than conventional CRT models. Flat-screen CRT TVs have reduced glare, but not necessarily the enhanced picture of a plasma or LCD display.

Comparing TV pictures? Compare contrast ratios. Contrast ratio is a measure of color representation (how the color information appears against a dark background). The higher the number, the better the color representation.

Source: Department of Energy, 2009; www.energystar.gov/index.cfm?c=home_elec_details.fap_tv_whatelse#Display

I understand how much people love their TV. I will not preach or recommend that you sacrifice your best friend—your TV, that is. What I do ask is that you reduce the number of monitors in your home and attempt to buy Energy Star models.

Now that you have a plan, you may have noticed that some of your plans you cannot or do not want to do yourself. Now you will require a knowledgeable and skilled contractor. So where can you find such a person? I'm glad you asked. Our next discussion is how to contract out your project. Contracting projects can be complicated for many reasons: size, scope, and cost are only a few reasons. Contracting can be difficult, but I hope to provide you with the tools to choose the correct person to create your dream home remodel.

CHAPTER 7

How and When to Contract Out Your Green Renovation

This chapter will provide you with an exceptional reference for contracting out your green home renovations. If you have decided that your home project is beyond your physical abilities to accomplish, you are probably correct in your assessment. If Figure 7-1 represents your typical home-improvement project, then contracting is for you.

Contracting out your work, however, does not mean that your job is complete. You are still the person in charge of your renovation. There is much more to contracting out your home remodel than just hiring a service provider. When you have completed this chapter, you will have the knowledge and the comfort to contract out your home project. Developing a superior and successful home remodel is a function of your planning, not your contractor's abilities. Your builder or designer can only bring to life your ideas. It is paramount for you to understand what you want to accomplish in great detail. You then must convey your vision to your contractor in great detail. If you completely understand only this single concept, your home remodel will be a success.

We Want to Hire a Professional

Let's begin with a quick checklist for hiring a remodeler; then we will go into more detail.

1. Know what you want, and be specific in explaining your needs and ideas.
2. Know what type of contractor you require.
3. Understand what information you must provide the contractor.

FIGURE 7-1 When to contract out. *(www.publicdomainpictures.net.)*

4. Interview the contractor, and know what to ask.
5. Understand what to expect from the first meeting with the remodeler.
6. Do not be afraid to ask for a second meeting.
7. Do your homework during all stages of the project.
8. Understand what you should expect from your remodeler.
9. Do not settle for what the contractor wants to tell you or sell you.
10. Do not change your plan unless you prefer the improved proposal more than the original.
11. Know your budget, and compare costs before making a financial commitment.
12. Explain to your contractor that you are in charge of the project.
13. Plan the project in its entirety, including life disruptions. Your project probably will go over your time limit. Unless you are leaving your home during the renovation, you will have personal and profession problems that interfere with the project timeline.
14. Build the team; be sure to use an architect, designer, or other expert when necessary.
15. Be sure to have all service providers approve your plan before work begins.
16. Obtain detailed plans from all service providers.
17. Personally approve all contractors or service providers, including subcontractors.
18. Do your homework. Check the contractor's references, as well as memberships, insurance, and licenses.
19. Manage the project, and check on progress frequently.
20. Do not be afraid to stop the project or fire the contractor at any time.

Let's take a short break and accomplish some immediate cost savings.

Appliances

Unplug Appliances When Not in Use

Unplug your home audio and video equipment when not in use; in fact, unplug all your appliances when not in use. Use power strips, and turn off all appliances plugged in at once. Vampire power can be as much as 15 percent of your electric bill.

Turn Off Your Computer

Turn off the computer. Turning off your computer is always more efficient than power-save mode, sleep mode, etc. Today's computers start fast and use very little energy in doing so.

Turn Off Your Dishwasher Before the Dry Cycle

Air dry your dishes even if you wash them by hand. However, the dishwasher dry cycle requires a tremendous amount of electricity. When the dishwasher completes the wash and rinse cycle, open the door of the dishwasher and pull out the drawers. The dishes are especially hot and will dry in minutes in the open air. No additional energy is required. Try this one, and you will never let your dishwasher go into the dry cycle again.

Use Your Large Appliances After 9:00 p.m.

Use your large appliances—washer, dryer, dishwasher, etc.—after 9:00 p.m. Most utility companies have lower rates in the late evening.

Unplug Your Power Adapters

Unplug your chargers and power adapters when not in use. Most chargers convert 120 volts ac to 12 volts dc. Those little black adapters are transformers. They will use energy when plugged in. Too lazy to unplug all your adapters? Get a long power strip (Figure 7-2), and just turn the power strip off when not in use. It is that simple.

Figure 7-2 Power strip with surge protector.

Use Appropriately Sized Appliances

Always use the right size appliance for the job. When I was a child, my mother would use the full size oven to heat English muffins. Needless to say, this involved the use of considerably more energy than was necessary. A toaster oven would be appropriate for this situation. This is just one of many possible examples.

Consult Your Family and/or Roommates

Deciding when and how to renovate your home is important. Planning, organizing, and sharing with your family or roommates (Figure 7-3) is the best approach to renovation. Include everyone who will be affected by the renovation in the decision-making process.

Planning Your Green Renovation

The most important part of any home remodel is the planning. Today, planning is made easy. Access to the Internet and a search engine can provide you with a bounty of valuable information. This information can include complete remodeling plans, prices, and contractor information.

FIGURE 7-3 Decisión making. *(http://www.cdc.gov/niosh/nas/mining/images/EmergencyCommunication.jpg.)*

Detailed product information is available in abundance. If you are choosing to use a green product such as paint that contains no volatile organic compounds (VOCs), you can choose brand, color, and type; compare by price; and be able to choose the exact product of your design dreams.

The first item to consider when planning your project is the timeline. Second, determine the complete scope of your project. Finally, establish your exact requirements. I will use a single-room remodel as an example. Develop your remodel in reverse. If you are remodeling a single room, understand how the room will be used when it is complete. This will assist you in designing the room of your dreams rather than one more area in your home. You will want to consider the appearance of the room as well as the function of the space. Everyone has a light switch in their home that is misplaced; it is okay to have a problem in an existing room. To build such a problem into a new design would irritate you every day for the rest of your life. Proper planning should identify every detail, no matter how small or seemingly insignificant. This will allow for better communication between you and your contractor during the remodel. The plan is what will provide you with a finished room that you will truly enjoy.

Use multiple communication tools when describing your remodel. Identify and choose your colors, patterns, and materials. Attempt to gather samples for your contractor. Draw your plans on paper or computer, including colors, furniture, lighting, entrances, windows, and doors. This will assist you in understanding whether your plans will fit your perceived needs. If your home addition is for a theater room and your scale modeling shows that your planned room is too small, it is easy to change now. If after you have created the theater and purchased the furniture you realize that you cannot fit your television in the room, you have made a costly mistake.

Do not be afraid to employ a professional. Most builders and homeowners do not use an architect. This is a mistake. An architect will cost you a few thousand dollars for most projects. The benefits of using an architect are numerous. You will be able to see the project in its entirety. The plan will be clear, concise, and detailed, and if the job is not completed to satisfaction, you now have the plans on which to base a triumphant lawsuit. Hiring a professional not only will save your money, but it also will reduce your frustration. You can find an architect in your area from the American Institute of Architects (AIA) Web site: www.aia.org.

Know and remain unyielding with your budget. Most projects go over budget by 20 percent. The expanded budget occurs because of unforeseen variables. Be sure to budget additional funds for potential problems. If you cannot afford the project you desire, change your plan, not your budget. Do not attempt to deceive yourself and assume that you will be able to reduce costs as your project progresses. Employ the assistance of a spouse, family member, or friend—someone who can say "No" to you. Review your plans with this person. Your "person of trust" will help to prevent you from making poor decisions.

You can negotiate with contractors. They do have variable costs, so you can get a better price—sometimes. The time to negotiate with your contractor is before you sign a contract and the work begins. If you attempt to negotiate after the work has begun, you will anger your contractor and probably entice him or her to do shoddy or incomplete work and leave your project as soon as possible.

Renegotiation because of unforeseen difficulties is a separate issue, and your contractor will understand. Even experts find things that they have never encountered, such as opening a wall and finding a cement or brick structure that requires additional work. Reconfiguration of the plan is common and acceptable under these circumstances. Losing your job may be an unforeseen difficulty to you, but it is not a reason to renegotiate with your contractor. Second, you may want to make changes to the original plan during your renovation. Changes equate to money. If you can afford and believe that you will enjoy the changes, *and* your contractor agrees, then go ahead and make the changes. If you have planned properly, this most likely will not occur. The final stage of planning is understanding how to investigate your contractor to obtain the desire outcome. I will provide you with the legitimate resources to do a proper investigation.

What Type of Contractor?

Know what type of contractor or contractors you require for your home-improvement project. You are the person in charge of your home remodel, not your contractor. I cannot emphasize this point enough. For this reason, you must choose the correct people to manage your job. You are responsible for building a successful team. Your team may include an architect, builder, contractor, handyman, helper, waste management person, and friends and family. You may use some, more, or all of these people. You may require only one or two of them depending on the scope of your project.

So what type of contractor do you need? This decision depends on the size, scope, and complexity of your project. If you are installing insulation, a local handyman may fit your needs. Adding a room to your home will require a general contractor. The general contractor is responsible for the complete job. He or she may require that you find specialty trades such as plumbers and electricians, or your contractor may supply all the subcontractors. You must be sure to ask or specify.

The type of contractor will vary according to the project that you wish to perform. New energy-efficient lighting may require an electrician but also may require someone to apply drywall and paint after the electrician has completed the new lighting. When choosing a contractor, attempt to choose a contractor whose specific trade is directly related to your project. Your handyman may be able to perform your new lighting and spackle and paint for one great price, but when your home burns to the ground afterward, your insurance company will tell you that it will not pay for your loss because of your choice of an "unqualified" worker. Always use a licensed, trade-specific contractor.

The best way to find a licensed professional is to first ask your friends and family for a reference. Ask if they have used a company and were happy with that company's performance. This is usually the best way to find a competent contractor. However, there is always a *but*—but be sure to do your homework. Your friend may have had a good experience with a contractor, but that does not mean that the contractor is licensed and insured. You still must do your own homework. Ask your friends and family about bad experiences. Why? Because you can avoid choosing those contractors from the phone book.

The last word about understanding what type of contractor you need is *green*. Unless your contractor came to your home in a tie-dyed shirt, drinking his daily wheat grass, and telling you how great the sixties were, he or she probably is not a green contractor. Most companies and contractors now claim to be environmentally friendly, but most are not. Companies are using the "green" title to promote the same old contracting business that they have always owned. This does not mean that the quality of the workmanship is poor, but rather that the person is probably not the foremost authority on green technology. Once again, you are left with the responsibility to do your homework and know what you are purchasing before you buy it.

What You Must Provide Your Contractor

The most important item you need to provide to your contractor is a clear vision of your dream. You must know in detail what you want to accomplish. You do not need to know how to accomplish your dream; that is the contractor's job. But you must know what you want and be able to convey that information to the contractor. Finally, you must have a firm budget. The contractor will be able to tell you if your budget is appropriate for the size of your endeavor.

One particularly significant sign of trouble when hiring a contractor is if he or she does not listen to you. If the contractor does not take the time to understand your vision, then how will he or she be able to provide you with your dream? If your contractor tells you, "Don't worry, I'll take care of it," what he or she is really telling you is that you are a nuisance and that he or she will do the job that he or she wants. If you simply have a bad feeling about a contractor, ask him or her to leave. You do not want to dread having a contractor whom you do not like in your home every day.

Interview the Contractor

It does not get any more fun than this. There are so many issues that will affect your home project. You want your prospective contractor to be friendly, trustworthy, and fair in price. The best contractors usually cost a little more but are always worth the price.

You will want to begin by explaining your dream or your vision to the contractor. Present your prepared plans, drawings, and specifications available to the contractor. Have any illustrations or product samples available. Providing the contractor with great detail increases the probability that he or she will perform to your expectations. Ask direct questions, and expect direct answers. If the contractor claims that he or she can perform the job required, then move on to serious questions about the contractor and his or her company. You must ask about insurance, years in business, licenses, and outstanding lawsuits. Do not be afraid to ask these questions because quality contractors will offer you this information without hesitation. Let's review each of these issues.

Make sure that your contractor is insured—with the correct insurance—and know the amount of his or her insurance. If your contractor is a plumber moonlighting as a general contractor and you have complications, his insurance may not pay your insurance claim or not pay it in full. If your project costs $2 million and the contractor is insured for $50,000, well, you can do the math. You must know all three things: Does the contractor have (1) insurance, (2) in the correct trade, and (3) for an ample amount?

Ask how many years the contactor has been in business. Having a business for many years does not guarantee that the contractor is reputable. It only means that he or she is still in business. If the contractor has been in business for many years, incorporated in the same name, he or she probably cares about his or her reputation. This will allow you to be able to do your own research.

The contractor should offer you a list of previous customers. This list will include names, addresses, and phone numbers. Former customers love to rave and display their grand vision brought to life by the contractor. Let me give you a personal example of a fantastic contractor. I needed to replace my roof. I was given a recommendation. I placed a call to the contractor, and the owner of the company came to my home at my convenience. The contractor gave me an exact price, not an estimate; he guaranteed his work and allowed me to choose from name-brand quality materials. Last, he gave me a list of over 100 customers within half a mile of my home. This contractor also gave me a list of 5 customers who were not immediately satisfied with his work. Three of the five people who had complaints did not like the color of their new roof. These people expected the roofer to buy new materials and pay for the labor to reshingle a brand-new roof. The other two complaints were from people with extremely complicated roofs involving coordinated projects and other contractors. Minor leakages had developed from construction that occurred by a secondary contractor after the roof had been completed. The roofer repaired these minor problems at no cost to the owner.

A contractor giving you a list of the people who filed complaints against his or her company is a confident contractor. He or she almost always will be nearly flawless in performance. The contractor should be completely confident in every aspect of the job, including how their product or service relates to the other elements of your home.

Be sure to check your contractor's license. Check that the license is current. Check that the license allows for the type of work sought. Confirm that the license is for the geographic region in which you are living. Ask if the contractor has any outstanding lawsuits. Contractors who have something to hide will always say no. Confident contractors will be honest and not make excuses.

Get three or more quotes, and explain that you are obtaining multiple quotes for each project. Bring contractors to your home one at a time. They may know each other, and you do not want the contractors to conspire. You also do not want squabbling or even playful competition. You want the contractor to focus on you, your needs, and your home.

Don't be afraid to ask the contractor to leave at any point in the estimation process or project. This is important. The farther you are into the remodel, the more difficult it will be to replace a contractor. In addition, no contractor will want to attempt to assume someone else's project. Attempt to dismiss contractors during the interview process. You are providing them with a good amount of your personal funding; do not be afraid to take charge.

Above all else, *get it in writing* (Figure 7-4)!

FIGURE 7-4 Contract. *(www.epa.gov/radon/images/consguide_contract.jpg.)*

Please attempt to spell out every detail of the project. If the project involves painting, specify the brand and type of paint. XYZ Paint is the best paint you can buy, but XYZ Paint does sell different grades of paint. Know your products, and understand what you are paying for. The more detail you specify, the better your contractor can understand and create your vision.

If you are unsure of some details, ask the contractor for a second meeting. Bring your spouse or a friend with you to the second interview. If you are unsure of the details, have the contractor explain the items until you are satisfied.

This is now the time to negotiate price. Now that you have obtained three or more quotes and you have good contractors to choose from, you can negotiate price. You always must negotiate price before the job begins. You can negotiate the work, materials, labor, and time. Do not attempt to negotiate after the job is started or completed. You will only enrage your contractor, and if the job is not complete, he or she will rush to complete the job and leave your home as soon as possible.

Ask how the work will be done. What do I mean by this? Well, it is your right to know that the contractor will work from 8:00 a.m. to 5:00 p.m. Monday through Friday. If you are expecting to sleep late, you must tell the contractor now. If you wish to expedite your project and want the contractor to work 12-hour days and on weekends, you must specify this before work begins. Remember, all details are important; do not assume anything.

Will your contractor subcontract any part of your job? This is important for quality, but also for insurance reasons. Do not assume that the subcontractors will be licensed and insured. Ask for a list of subcontractors, and ask to interview them. Ask all your contractors if they have performed this type of work before. Ask your contractor if he or she has used these subcontractors previously.

What You Should Expect from Your Contractor

This is very simple if you have interviewed contractors, explained your home remodel in detail, and then chosen your contractor. You now possess a detailed plan and written contract, so you should have no questions in your mind about your contractor or the work that he or she is to perform. If you still have questions, you need to review what you have accomplished and have your questions answered before work begins. A second interview or meeting to finalize details is very common. Do not be afraid to call the contractor back into your home to finalize the details.

The First Interview

- Most important are the questions from the contractor. Do not panic. He or she was listening to you. Remember, it is your dream remodel, not his or hers. He or she will need clarification.
- The contractor will want to see complete, detailed plans.
- The contractor should be able to provide you with a list of the necessary permits. If the project was too large for you to do, the permit process is out of your league.
- The contractor will know the cost of your project and/or products.
- The contractor will provide proof of insurance.

- The contractor should provide you with samples of products and materials and references to jobs completed.
- The contractor should provide a detailed estimate. Know the cost of your project; at least then you will not pay too much. Always ask the contractor for a detailed estimate, itemizing each item. Then you can do a fair comparison of estimates. You also can go to the local home center and check the prices for yourself. Keep in mind that the contractor will employ higher prices for the products used. It does require his or her time to shop, buy, and deliver the products to your home and then wait for you to pay for them.
- Do not write any checks until you are satisfied with the written contract.

Never pay a contractor up front. Most legitimate contractors will not ask you for any money until the job is complete. There are some exceptions. Small contractors will ask you for one-third of the total cost for materials and initial labor, a second payment of one-third after he or she has completed most of the job, and the final payment when the job is complete. This type of contractor does not have large purchasing power.

The only other legitimate reason you will be asked for funding prior to the start of a project is if you have specified custom materials. The contractor will want the price of all custom colors, materials, appliances, etc. prior to beginning the project. The custom products cannot be returned for full price, and your contractor does not want to become the proud owner of a new lime green shower and bath. If you are uncomfortable giving the contractor funding prior to the start of the project, you probably have selected the wrong contractor.

As I have stated, allow the contractor to obtain the building permits. If you are contracting out a home project because it is out of the realm of your capabilities, so is the permit process. Improper permitting could cause significant delays or noncompliance with the building code. This will cause delays or require you to pay additional money to change or complete the project. Let your contractor obtain the permits.

Building codes are established at all levels of government. Figure 7-5 shows the building code requirements for insulation. Building codes are implemented for your safety and your benefit. You may not understand or agree with the building codes in your area, but I assure you that the permit process is in place for your safety. Building codes are, for the most part, a process of evolution. Tragedy is usually the precursor to new building codes. You may have noticed that the structures in most cities are built of stone, concrete, and steel; most suburbs are built of wood; and rural areas have a mix of materials. The requirements have been dictated by the vulnerability presented by the circumstances.

Zone	Gas	Heat pump	Fuel oil	Electric furnace	Ceiling: Attic	Ceiling: Cathedral	Wall (A)	Floor	Crawl space (B)	Slab edge	Basement: Interior	Basement: Exterior
1	✔	✔	✔		R-49	R-38	R-18	R-25	R-19	R-8	R-11	R-10
1				✔	R-49	R-60	R-28	R-25	R-19	R-8	R-19	R-15
2	✔	✔	✔		R-49	R-38	R-18	R-25	R-19	R-8	R-11	R-10
2				✔	R-49	R-38	R-22	R-25	R-19	R-8	R-19	R-15
3	✔	✔	✔	✔	R-49	R-38	R-18	R-25	R-19	R-8	R-11	R-10
4	✔	✔	✔		R-38	R-38	R-13	R-13	R-19	R-4	R-11	R-4
4				✔	R-49	R-38	R-18	R-25	R-19	R-8	R-11	R-10
5	✔				R-38	R-30	R-13	R-11	R-13	R-4	R-11	R-4
5		✔	✔		R-38	R-38	R-13	R-13	R-19	R-4	R-11	R-4
5				✔	R-49	R-38	R-18	R-25	R-19	R-8	R-11	R-10
6	✔				R-22	R-22	R-11	R-11	R-11	(C)	R-11	R-4
6		✔	✔		R-38	R-30	R-13	R-11	R-13	R-4	R-11	R-4
6				✔	R-49	R-38	R-18	R-25	R-19	R-8	R-11	R-10

FIGURE 7-5 Insulation requirements. *(www1.eere.energy.gov/consumer/tips/images/chart_new_construction.gif.)*

If Mrs. O'Leary's cow kicks over the lantern in your country barn, your barn burns to the ground. If the same bovine kicks over your lantern in Chicago, when the city is made of wood, the entire city burns. In low-level areas, flood requirements are in place to protect you. In a hurricane or typhoon zone, building requirements are necessary for high wind and debris protection. In Japan and California, in the earthquake zones, buildings are also built accordingly. To ignore your local building codes is to place the life of you and your family in jeopardy. Please think about this before you attempt to save a little money and time by not obtaining the correct permits.

If the preceding rationale is not reason enough for you to obtain the proper building permits, then how about money? If you do renovations to your home and you do not obtain the proper permits, when you attempt to sell your home, you will not be able to do so. During the sale of your home, you will need to obtain the proper permits and possibly redo the work originally completed without the permits.

Zoning and permitting laws vary considerably from one jurisdiction to another. Be sure that your professional is familiar with the process of obtaining permits. Your contract should state that the work performed will be in accordance with the applicable building codes in your area. A building permit typically is required when you are performing structural work to your home. Special trades also require permits. Electrical, heating and cooling, and plumbing are all trades that require special training and special permitting for your home. Licensing and permitting are almost always approved for average projects because your local town will want to assist you in improving your home. Your home improvements theoretically better your neighborhood and therefore your town's tax base.

Your contractor should apply for the permits via his or her company. This allows for your financial and legal protection. If the contractor's work does not pass inspection, you cannot be held responsible financially for the corrections that will need to be made. All issues and items should be negotiated with the contractor before work begins and put into writing in the form of the contract.

When the work is almost complete, your contractor should arrange for your town to inspect the completed work. The inspector will come to your home and verify that the work performed complies with the local building code.

Be sure to have your contractor obtain all the permits and variances that affect your project. If you are not familiar with the building codes, ask if your contractor is capable of obtaining the proper information and permits. If not, he or she probably does not do large-scale renovations. Have your contractor provide you with copies of all permits.

Finally, your contractor may wish to be compensated for the interview time. If you have an extensive remodel and your contractor will be spending days assisting you to develop a plan, expect to compensate him or her for that time. This is an acceptable payment.

Contract Negotiations

Perhaps the greatest mistake is to choose a contractor because of a low price. Do *not* do this! I will reiterate this important item many times. Contract negotiations occur before the project begins, and all details are put in writing. Negotiations do not occur after the project has started or after the project is complete. Any variation will cause difficulty. There are some exceptions to this rule, but I will discuss them later.

Negotiation of the contract should be as specific as possible. All terms and conditions need to be specified in writing and agreed to by both parties before work begins. A written contract will protect both the client and the contractor and allow the project to progress smoothly. If you intend to do any of the work yourself or hire additional assistance, such terms should be understood by your contractor and put into the contract.

The written contract also should include:

- A detailed description of all work to be completed
- Start and completion dates and consequences if the dates are not met
- An itemized inventory of materials
- Specified quality and quantity of materials to be used
- Identified colors, brands, and sizes
- Documented itemized inventory of all costs, including labor
- Created and agreed-on schedule for payment
- A detailed plan for debris removal
- Guarantees and warranties for both the contractor and the materials
- The right of refusal—ability to cancel the contract before work begins or material is purchased
- The right to cancel or renegotiate the contract if unforeseen problems arise

A standard contract should include all the contractor's information, all the client's information, license numbers, insurance information, and the date at the time of signature. The contractor will have a standard contract form. Ask him or her to explain each part, including the preprinted small type. Do not sign a blank contract after a verbal agreement with the contractor. You should receive and retain an identical copy of the signed contract for your personal records.

After the contract is signed, then payments may be disbursed. Most contractors who follow the informal rule of the one-third, one-third, and (guess what) one-third payment plan will expect a check or cash for the first third of the amount of the job. This is an unwritten rule, and many firms of significant size and expertise will not ask you for any money until the job is complete. These firms are confident that the materials they use and the workmanship employed will allow for a flawless performance. I recommend that you find such a contracting firm for your projects. These firms typically cost about 10 percent more than the market average, but I have found that the job usually is completed on schedule and frequently exceeds expectations. This is what you want for your home project.

If you have chosen the typical one-third rule for payment and you have not done business with the contractor previously, you should attempt to pay no or a low down payment until the contractor has begun work. You should not pay the contractor if he or she is behind schedule. Payment should be made for phases of completion, not time intervals.

Payments should not be made for incomplete or poor-quality completed work. All these items should be identified clearly in the contract.

In most states, you have the right to cancel your contract within 3 days of the initial signing. Each state or country has its own rules and regulations. Be knowledgeable before you sign the contract. The contractor should inform you of the right to cancel. He or she also should explain the process of cancellation. If the contractor does not inform you, ask what your rights are according to the contract.

Never provide the final payment or release until you are satisfied with the completed work. Verify that all subcontactors have been paid. If the subcontractors or supplies are not paid by the contractor, you may be liable. You can completely avoid this problem with proper planning. As the client, you can require a "release-of-lien" clause in the contract. Payments to subcontractors can be placed in escrow and not disbursed until work is completed to your satisfaction and to the satisfaction of your general contractor.

Warranties of materials or workmanship should be in writing. Careful attention to detail will prevent any misunderstandings. If you do not understand, you must ask before signing the contract. Understand all terms and conditions of the warranties. Recognize the terminology; for example, what is a limited lifetime warranty? Be familiar with the length of time, physical limitations, geographic locations, and your rights as the consumer of the product and/or service. If your remodel is because of a natural disaster or accident, be sure that your contractor will work with your insurance company or within your insurance budget.

Do Your Homework

If you take the simple steps that I have recommended and you do your homework, you will be satisfied with the outcome of your home remodel. If you are driving to another country, do you just get in your car and go? No, you research your path, you plan for your expenses, and you plan the timeline of the trip. This is exactly the same procedure you should follow for your home project or remodel.

There are many tools to investigate your contractor; the best and most trouble-free tool is the Internet. You can use the Internet to check with local agencies and consumer affairs groups and even speak directly with consumers. The following list is a good beginning for your investigation:

- Check that the contractor's license and insurance are valid and up-to-date.

- Be sure that the contractor's insurance is complete and does not leave you liable. Important items include the amount of the insurance, the type of damage, and worker's compensation.
- Check with the Better Business Bureau (www.bbb.org). You are looking for unresolved complaints, numerous complaints, and fines; these are all indicators of a bad contractor.
- Check with private building associations.
- The National Association of the Remodeling Industry (NARI) can give you a private review of your contractor (www.NARI.org).

What questions should you ask of these organizations? When you do not know what to ask, ask the organizations themselves. These associations are extremely helpful. This short list of inquiries also will assist you:

- Ask about the contractor's history.
- Examine the contractor's success rate based on completion, schedule, cleanliness, and work habits.
- Ask general questions of previous clients, including
  - What did you think of the contractor?
  - How satisfied were you?
  - Did the contractor meet or exceed your hopes?
  - How satisfied were you with the quality, professionalism, and final result?
  - How easy was the contractor to work with?
  - How or were any subcontractors used?
  - Would you use this contractor again? (The person you are asking must reply positively and immediately. If the person hesitates or does not appear overjoyed about the work done, choose another contractor.)
  - Was the contractor available to answer question during the remodel?
  - Did the contractor keep information about the project and its status?
  - Did you have to change plans, and was the contractor difficult to work with during this process?

When to Say No

This is very simple to know but difficult to do. Ask the contractor to leave at any time you think it is necessary—if you do not like your contractor's attitude, if you have a bad feeling, if his or her explanations or lack thereof are not satisfactory. Anytime that you do not feel in control and cannot take control of the situation, it is time for you to remove the contractor. You will recognize when you should remove the contractor, but most people still hesitate. The moment anything begins to go wrong, stop the project and resolve the situation before it becomes necessary to remove the contractor.

The Contract and Your Protection

A contract is an agreement between two or more parties. The type of a contract for a remodel job is a *performance agreement*. You agree to pay the contractor a specific fee for him or her to perform in a certain manner. Once the contract is signed, each party is expected to perform as required by the contract. The contractor has a legal document to protect him or her. Read that contract and change what you do not like before you sign it. The contractor has a brief preprinted contract. This contract can and should be altered to create a custom contract that meets your requirements.

During your final meeting with the contractor, both parties will sit together and edit the contract as needed. You can remove preprinted items, and both parties should initial any changes to confirm that the contract has been edited and agreed to by both you and the contractor. Most preprinted contracts have space for additional requests. This is where you write in your specific requirements. Then both parties sign and date the contract. Both parties also receive a copy of the contract. The contract is now legally binding, with the exception of the 3-day rule. Most contracts allow the client to terminate the contract for any reason within 3 days of signing. This is to allow homeowners time to consider their responsibility.

Most quality contractors also will allow you to terminate the contract if you wish to make major changes to the contract. Both parties will meet, agree to terminate the first agreement, and create a second agreement.

What to Do When Things Go Bad

The best way to avoid problems is to do what I have recommended all along. Inexorably, things do go wrong in life. The best way to avoid major issues is with proper planning. When problems do occur, resolve the issue ASAP. Resolution of problems with the contractor is the most expeditious and cost-effective solution available to you. Any other alternative will cost you significant amounts of time and money.

Before confronting your contractor about problems, gather some information. If your contractor or his or her workers begin to show up late, be cordial, and inquire as to why he or she is not acting as promised. Before reprimanding, understand that your contractor does more than just work. Everyone has life disruptions, so be understanding. Your contractor may be having sudden life problems. A death in the family, a birth, a car accident, etc.—these are issues that affect one's life in totality. If the contractor is showing up late because he is on another job, you have a right to be angry and to attempt to correct the problem. Identify and deal with problems or conflict immediately. Address all issues before they become a significant cost or time concern.

If you are reading this portion of the book, and you are having problems with your contractor, it is probably too late for you and your remodel. You are the unhappy homeowner with an uncompleted project. There are too many reasons why home projects fail, including everything from unrealistic homeowner expectations to direct fraud by contractors.

What is important at this time is resolution with the contractor. If you are still speaking with your contractor, attempt to deal directly with him or her to complete or redo the project for a reasonable fee. This negotiation is usually the least expensive and painful option for the homeowner. If you cannot speak to the contractor, begin by filing a complaint with the Better Business Bureau and your state consumer complaint board. These organizations may contact the contractor and attempt to act as an intermediary to resolve the issue. You will not have to speak to the contractor directly.

Your final option for resolution is to obtain legal advice from an attorney. I am sorry, but how to choose an attorney is a significant process that this book could not possibly address.

All the References You Will Ever Need from Start to Finish

Your best and most unbiased resources are always free. Your state consumer protection agency, the Better Business Bureau, and the National Association of the Remodeling Industry (NARI) will assist you for free. These organizations can provide you with a list of contractors in your area, the contractors' histories, claims filed against them, and often the resolution.

NARI represents remodeling contractors, product manufacturers, and other industry specialists. The association is committed to enhancing the professionalism of the remodeling industry and serving as an ally to homeowners. NARI members must abide by the NARI code of ethics. NARI certification includes exams and certification classes. NARI does require a membership fee. This may appear to be a conflict of interest, but in my experience, NARI is reputable, honest, and accurate in its evaluations of contractors. For more information about NARI, visit NARI.org.

Other resources and channels for consumer complaints include:

- Your Local Better Business Bureau at www.BBB.org
- The Federal Trade Commission at www.FTC.gov or 1-877-ftc-help (1-877-382-4357)
- National Association of the Remodeling Industry (NARI) at www.NARI.org or 1-703-575-1100
- National Association of Home Builders (NAHB) at www.nahb.org or 1-202-822-0216
- National Association of Realtors at www.realtor.org or 1-202-383-1000

When filing a complaint, you will begin with your state office of consumer protection. You can fill out the forms online, print them, and mail them or file a complaint over the phone. The Office of Consumer Protection will contact the contractor directly and attempt to resolve the complaint without the need for you to speak directly with the contractor.

If there has been fraud or legal action is necessary, you can contact your state attorney general's office, and once again, that office will contact the contractor directly. You can find these resources online or in the blue pages of your phone book. Most organizations have an online claims form making the process simple. When filing a complaint, state the facts, and leave out emotion or examples. These organizations care only about the law.

Your last venue to attempt recovery on a poorly performed project is court. If you are seeking small retribution, then small claims court is for you. The process is simple and straightforward, and a judge makes the decision immediately. If your remodel was expensive and you cannot resolve the issue in this manner, your last alternative is to take the contractor to superior court. You will require an attorney, and you will need to document everything that occurred.

Most attorneys will have an initial phone or in-person consultation with you for free. The attorney will determine if you have a viable case for court. Even though you may have been wronged, this does not necessarily mean that you can be compensated financially by the court. How will you know this? Attorneys will take a case only if they can benefit financially. You also may be able to recover court costs. Speak to an attorney in your state or county to be sure.

Home remodeling is your responsibility. If you choose to be an active participant, then the probability of a successful job is great. The burden of success falls on you. Of course, if the final creation of your home remodel is greater than your wildest expectation, you can take credit for its success. The choice is yours. Now that you know how to contract out your project, let's take a look at large home remodels.

CHAPTER 8
Large Renovation Projects

Large is a relative term. For some individuals, the term *large* may mean opening windows; for others, installing a window; and for still other individuals, large may mean the ability to build an addition to a home. No assumptions will be made about *ability*. Large will be what you, the homeowner, are comfortable accomplishing.

Now that you know how to hire a contractor, let's discuss the reasons why you should or should not hire a contractor. Contractors are expensive. Approximately half the money paid to a contractor goes directly to labor; the rest is for materials. Completing a home remodeling project yourself can save you money, provided that no mistakes are made and that the project can be completed in a reasonable amount of time. The home-improvement project that will be thoroughly demonstrated here is the installation of insulation. This is a project that can be accomplished by the homeowner or by a contractor. The limiting factors will be the home design and the materials and tools required to complete the installation.

The goal is to improve the energy efficiency of the home. Lack of insulation is the single most prevalent problem found in residential homes. I will take you through the process of accomplishing a large renovation by creating a total home solution. We will create together a complete home energy envelope. The energy envelope will be created with insulation in many forms, including doors, windows, weather stripping, and all the major sources of air transference. The energy envelope is the most important home efficiency item. The energy envelope must be treated as one item.

The plan will include all that you have learned about hiring a contractor. Since the energy envelope is a large remodel, you may choose to use a(n):

- Architect
- General contractor
- Builder
- Specialty contractor
- Handyman
- Friends and family

The people selected to perform the project will be chosen based on your specific requirements.

Large Renovation Projects

By now, you should have a good grasp of the skills both to plan a remodel and to hire a contractor. Therefore, I will move directly to the project. Creating a home energy envelope is a macro assignment (Figure 8-1). The most obvious way to view problems or changes is from the outside of your home.

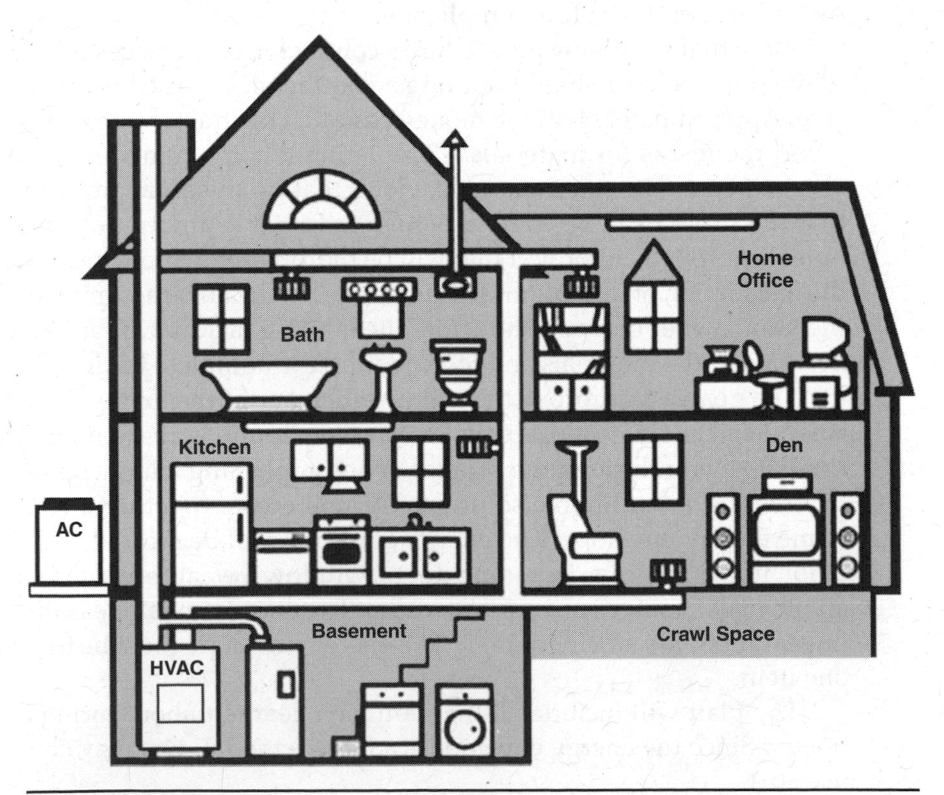

Figure 8-1 Macro view of a house. *(www.epa.gov/iaq/homes/hip-front.html.)*

Viewing your home and creating a diagram such as that shown in the figure will help you to identify the areas that are in need of improvement. By viewing the home from the outside, you should see the larger needs of the house immediately. Windows and doors and attic and foundation insulation all demand immediate examination. Make a list of known and suspected items that need improvement, and then move indoors. Inside, confirm the status of the suspected items. Check the doors and windows, specifically looking at the weather stripping. Look in the attic, and measure the amount of insulation. Confirm all the suspected areas that may be exchanging air with the outside.

Compare your findings with the information in Figure 8-2. Then continue with the major items on the graph. According to the U.S. Department of Energy (DOE), warm air leaking from the home during the winter and cool air leaking out of the home during the summer are the primary sources of wasted energy in the home. This energy loss therefore is the basis of the homeowner's primary financial loss through the home. Luckily for homeowners, air leakage is a problem that is also one of the least expensive and easy to remedy. Insulation, weather stripping, and caulking can reduce the flow of air from the home dramatically.

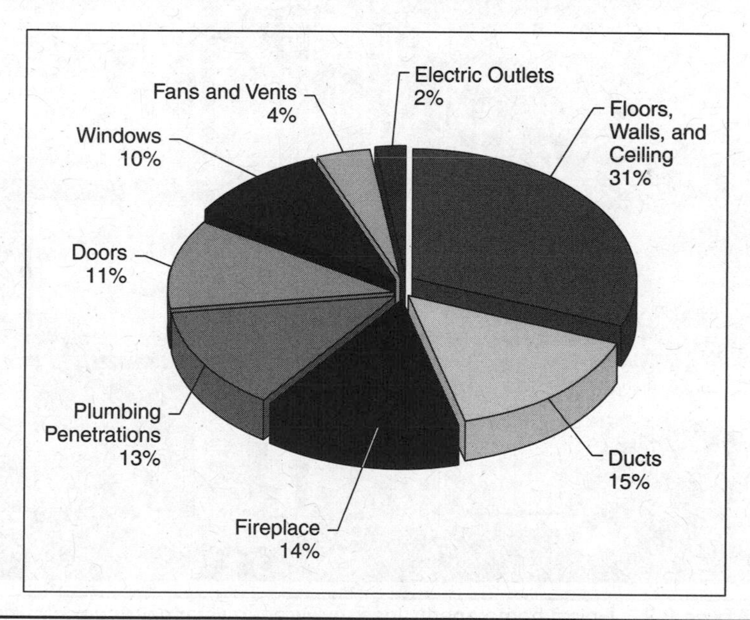

FIGURE 8-2 Energy losses. *(www.epa.gov/iaq/homes/hip-front.html.)*

Before air leaks can be sealed, they must be found. Air can and will leak from a home from anyplace that is accessible to the outside air. Air leakage occurs for two reasons: the temperature differential and the pressure differential. During the cool months, warm air in the home attempts to move to cool air outside. The second factor, pressure, is also at work. A warm home in the winter is in reality at a higher pressure than the outdoors. The pressure differential will cause the air to move from inside to outside in an attempt to equalize the pressure.

Think of a home as a water balloon. Any breach in the balloon will cause water to escape. The difference is that in a home, air leakage is not obvious unless you know where to look. Study the utility bills for the home, and understand that much of the money paid is for "leaking water."

Air does not actually *escape* from a home. The property of the air changes as the air enters and leaves the home. A home will lose heating, cooling, and moisture. Air can enter into and out of a home through every hole or opening. About a third of the air in the home is circulated from the floors and walls through the ceiling (Figure 8-3).

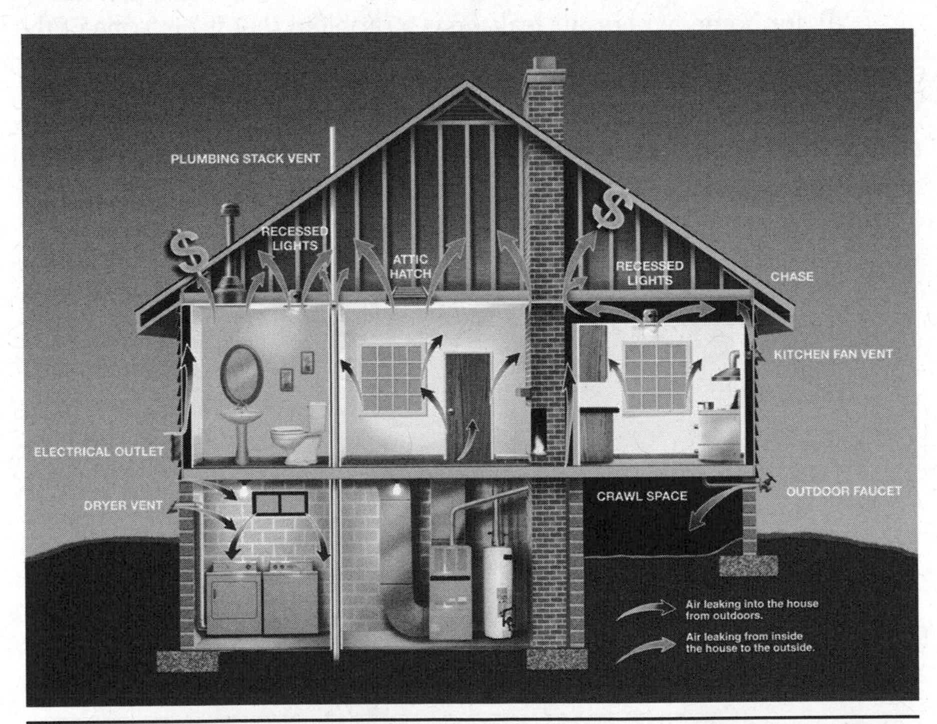

FIGURE 8-3 Typical home energy loss. *(www.energystar.gov/index.cfm?c=behind_the_walls.btw_airsealing.)*

There is virtually an unlimited number of areas where air can escape from a home. Existing homes are difficult to diagnose because modifications to the home have occurred almost continuously over time. A trick that the DOE recommends is to test the home for air tightness. On a windy day, carefully hold a lighted incense stick or a smoke pen next to your windows, doors, electrical boxes, plumbing fixtures, ceiling fixtures, attic hatches, and other locations where there is a possible air path to the outside. If the smoke stream travels horizontally, you have located an air leak that may need caulking, sealing, or weather stripping.

This method is good if the home is in need of obvious improvements. Homes that are more energy efficient and not drafty can be more difficult to diagnose. For a more precise identification of the energy envelope, hire a home energy audit specialist (Figure 8-4). The specialist mounts a blower fan in a doorway and closes all other exits and windows. This device will provide an accurate measurement of how much air is being circulated through the home. I recommend that this test be done at the beginning of the audit and after repairs have been completed. A significant reduction in airflow should be noticed. The second test also will provide you with a good idea of where to look for the next areas of improvement. The DOE lists 11 common places where air leakage can occur (Table 8-1).

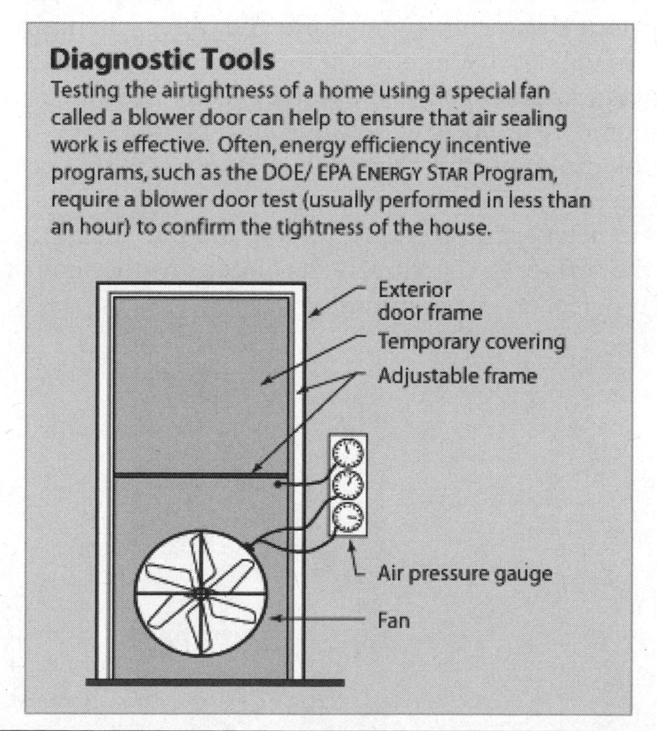

FIGURE 8-4 Air leak test. *(http://www.energysavers.gov/images/diagnostic_tools.gif.)*

Let's take a short break and accomplish some immediate cost savings.

TABLE 8-1 Sources of Air Leaks in Your Home

Areas that leak air into and out of your home cost you lots of money. Check the areas listed below.

Dropped ceilings	Water and furnace flues	Window frames
Recessed lights	All ducts	Electrical outlets and switches
Attic entrance	Door frames	Plumbing and utility access
Sill plates	Chimney flashing	

Source: Department of Energy, 2009; www1.eere.energy.gov/consumer/tips/air_leaks.html.

Heating

Use Your Resources Completely

During the cold months, close the drain when you take a shower. Let the water remain in the tub. This will cause the tub to become as warm as the water. When you complete your shower, open the drain. The water will drain, but this will leave a nice warm tub that will continue to heat your bathroom after the water is gone. You are using the hot water to become clean *and* to add heat to your home.

You can confirm the value of your new heat sink or tub with a point-and-shoot thermometer (Figure 8-5). The operation is simple: Point the thermometer at the object in question, and pull the trigger. This will show you the surface temperature of the object. The thermometer works via infrared detection and is accurate.

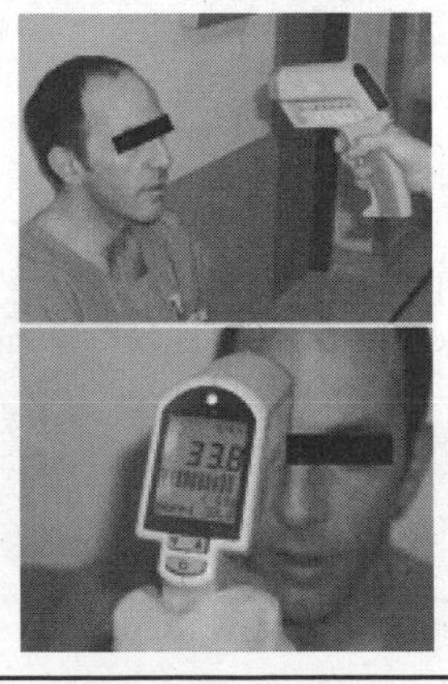

FIGURE 8-5 A point-and-shoot thermometer. *(www.cdc.gov/eid/content/14/8/images/08-0059_1t.gif.)*

Turn Off Your Bathroom Ventilation

During cold months, when using the shower or bath, shut off the bathroom fan to keep hot air and moisture in the home. You do not want to be removing all the heat from the room to the outside. Not only does this make your shower uncomfortable, but it also wastes heat energy. When you have completed your shower or bath, do not turn the fan on to defog mirrors, etc.; instead, open the door to the bathroom. The room heat and moisture will spread to the rest of the house and make it more comfortable, while probably defogging your mirrors more rapidly than the bath fan. Many older light or fan fixtures have only one switch. Upgrade to a new, more efficient fan with a separate switch or, for existing fans, separate the light and fan with a second switch. Both are cost-effective and easy to do. By taking these low-cost actions, you save heat and electricity.

Lock Your Windows

Lock your windows. This is a good tip for both safety and efficiency. During the winter, the windows contract because of the low humidity and cold weather. Windows can slowly slide open over time. For the windows to function efficiently, they must be pressed firmly against the weather stripping (Figure 8-6).

Review Placement of Your Heating System

Be sure that your heating system, including wood heat, is located in the central area of your home to allow for maximum efficiency.

Open Blinds in the Morning and Close Them at Night

Open your blinds in the morning, and close them in the evening. The light and heat are delivered free. Why not use them?

Run and Maintain Your Furnace or Boiler Properly

Have a professional clean and tune your furnace or boiler each year. Perform the maintenance just prior to the heating season. This will allow for maximum performance

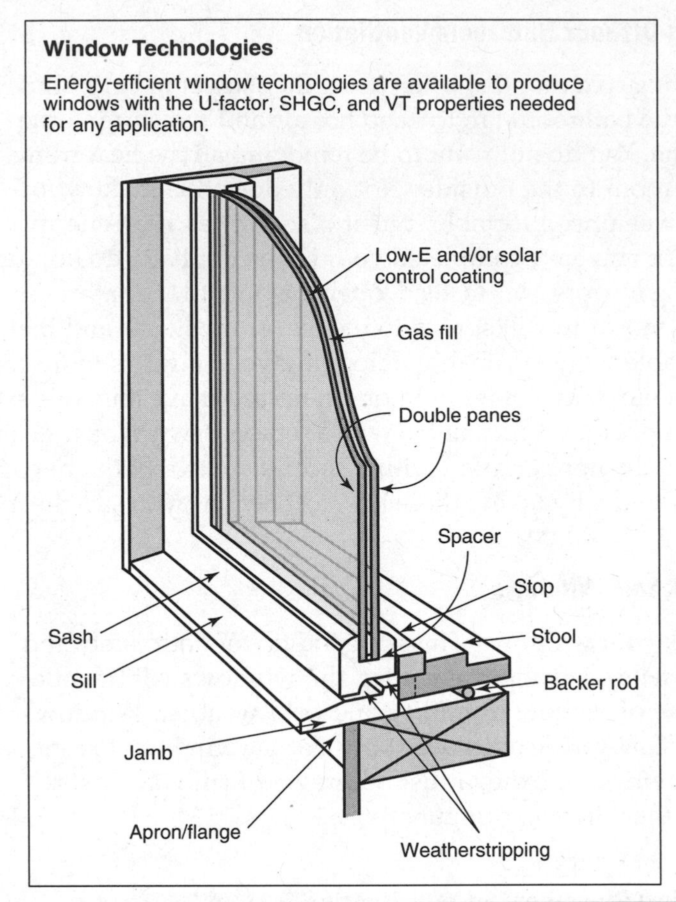

FIGURE 8-6 Window construction. *(www.energysavers.gov/images/window_technologies.gif.)*

when required most. In addition, check your furnace or boiler settings, and ask your service professional about anything that may affect performance.

Turn down the heat in your home 2 degrees in winter and up 3 degrees in summer. This small change in temperature probably will not be noticeable in terms of comfort but will save you money. In the colder months, it is often a lack of humidity that will make you feel cool. In the summer, it is excessive humidity that makes you feel warm. Be sure of the problem that you are addressing.

How do I know the difference? If your heat turns on often and you still feel cold, your home is not well insulated. The more that the heat is turned on, the more the heat and moisture are being removed from your house, and the cooler you will feel. Add a humidifier to your house. If you feel warmer, then you need more insulation.

Close Your Chimney Flu and Vents When Not in Use

Although it is good to use your fireplace during cool but not cold weather, at other times the fireplace can draw out more heat than it produces. Also, be sure to close the flu to eliminate drafts when the fireplace is not in use. In addition, keep air vents clear to allow air movement into your rooms, but close your vents when airflow is not needed.

Inspect and Repair Insulation

Inspect your air duct insulation for leakage and repair it before the start of the heating season. Insulate your attic, walls, and foundation properly to keep your house warm.

How Insulation Works

Your home's insulation will provide you with a thermal barrier. An added benefit to most forms of insulation is that the home also will have improved acoustical properties. Noise reduction is an added benefit of thermal insulation. A vapor barrier also may be a component of thermal insulation.

Most people know what insulation is by sight, but what is insulation really, and how does it work? The most common form of residential insulation is fiberglass batt insulation (Figure 8-7). We know that glass conducts heat and cold rapidly, so why or how is fiberglass insulation (strands of glass woven together) used as an insulator? A product that inhibits the transfer of heat or cold is insulation. This is simple. It is not the fiberglass that insulates; it is the air that is trapped between the strands of glass that provides the insulation properties.

Think of the blanket on your bed. It is a cold winter evening, and you want to be snug and warm in your bed. The blanket is made of many strands of thread that are woven together. Those strands of fiber create air pockets that resist the flow of air away from you when you are underneath the blanket. This is why you are cold at first when you crawl underneath your blanket but soon become warm as the blanket traps the heat that you are producing underneath the blanket. Remember our talk about conduction, convection, and radiation from Chapter 1? These are the ways that heat transfer occurs within your home.

FIGURE 8-7 Insulation. *(www.energystar.gov/index.cfm?c=behind_the_walls.btw_airsealing.)*

Types of Insulation

There are many types of insulation. All types, however, function in the same way—by trapping and limiting airflow. Today, homeowners have many options and types of insulation to choose from. Whether you are insulating a new home or retrofitting an older home, every home can benefit from more insulation.

Insulation can be made from a variety of materials. Each type of insulation has different characteristics. The most common forms of insulation are as follows:

Rolls or batts of insulation. This is the most common form of insulation and is made from fiberglass or rock wool (Figure 8-8). Fiberglass insulation is inexpensive and available in many sizes. A roll of insulation is one continuous roll of fiberglass. Batts of fiberglass insulation are cut to predetermined lengths.

Fiberglass insulation is available in widths suited to standard wall stud spacing and attic or floor joist spacing. Common insulation products come in R-13 or R-15 batts and R-19, R-21, R-30, and R-38 rolls. Insulation can be stacked to achieve higher R values. Do not compress any type of insulation. Remember, the insulation does not do the insulating. The air trapped by the insulation is what insulates. Compressing the insulation reduces the amount of air trapped by the fiberglass and thus reduces the R value.

FIGURE 8-8 Fiberglass insulation.

Loose-fill insulation. This type of insulation is made of many different materials today. These materials include but are not limited to fiberglass, rock wool, cellulose, and old bluejeans, to name a few. Loose-fill insulation is blown into spaces using a pneumatic pump (Figure 8-9). The blown-in insulation must be installed in a confined area. The material is blown in, and the cavity is filled and then sealed. Blown-in insulation is excellent for existing buildings, where it is difficult to install traditional insulation. The cost of loose-fill insulation is compatible with that of fiberglass batts.

FIGURE 8-9 Blown-in insulation. *(http://www.neo.ne.gov/neq_online/may2006/images/loosefill.jpg.)*

Spray-foam insulation. This type of insulation is becoming very popular. It is sprayed on a building's interior strtucture as a liquid and immediately expands into foam that fills the desired cavity (Figure 8-10). The foam dries and becomes rigid.

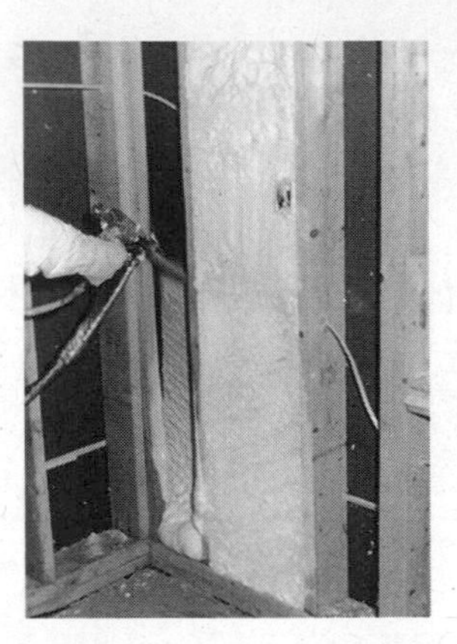

FIGURE 8-10 Spray-foam insulation. *(www.energystar.gov/index.cfm?c=behind_the_walls.btw_airsealing.)*

The insulating factor of foam insulation is almost double that of batt insulation. In addition, spray-foam insulation comes in open or closed cell form. Open cell insulation can be sprayed under the roof in an attic. Moisture will not accumulate, and condensation will not be able to occur. Open cell spray-foam insulation is resistant to moisture and allows for the transference of moisture. This transference is allowed because each foam cell is slightly open. The final benefit is that the foam expands to fill all holes, creating a true energy envelope. Closed cell spray-foam insulation is formed with completely closed foam cells. This allows for better insulation properties but will not allow for the removal of moisture.

Spray-foam insulation is more difficult to install, and it costs more than traditional batt insulation. However, spray-foam insulation is far superior in insulating ability (Figure 8-11).

Rigid foam-board insulation. This is the final type of commercially available insulation (Figure 8-12). Rigid foam-board insulation is denser than fiberglass insulation and therefore more efficient. Foam-board insulation costs more but is an excellent alternative when the insulation will come in contact with the elements.

Determine what types of insulation are presently in your home. If you are unsure, have a professional survey your home. I have seen a few homes with different types of asbestos insulation. This is a product that

FIGURE 8-11 Foamed holes. *(www.energystar.gov/index.cfm?c=behind_the_walls.btw_airsealing.)*

FIGURE 8-12 Rigid foam-board foundation insulation. *(http://resourcecenter.pnl.gov/cocoon/morf/ResourceCenter/dbimages/full/97.jpg.)*

no homeowner should touch. Asbestos happens to be a fantastic insulator. The problem with asbestos is that the fibers are practically indestructible. These asbestos fibers can be inhaled, become lodged in the lungs, and become covered with protective growths. This is called *asbestosis* and is a very serious condition that can lead to lung cancer. If you have asbestos of any type in your home, have a certified professional remediate the problem (Figure 8-13).

First, as mentioned earlier, identify the types of insulation that are in your home. Second, determine the amount of insulation and the R value. Third, determine if the old insulation should be removed or if new insulation can be added over the existing insulation. I will show you how to do all of these things and much more.

FIGURE 8-13 Traditional fiberglass insulation can cause health problems similar to asbestos. *(http://www.osha.gov/SLTC/etools/hurricane/images/photos/building-assessment/repair00.jpg.)*

The home survey should reveal at least two types of insulation. Each area of the home may use a different type of insulation. Check the insulation in your attic or ceilings. Check the exterior walls and the home's sill. Check the basement walls and any crawl spaces to see if the insulation there meets the R values recommended for your geographic area. Crawl spaces are notorious for having no insulation at all.

Once the insulation survey is complete, determine the amount of insulation required. As mentioned, insulation effectiveness is measured in R values. The higher the R value, the greater are the insulating properties. The R value is the *thermal resistance*. The DOE and large corporations such as Owens Corning provide excellent information about the R value required for your home based on location (Figure 8-14).

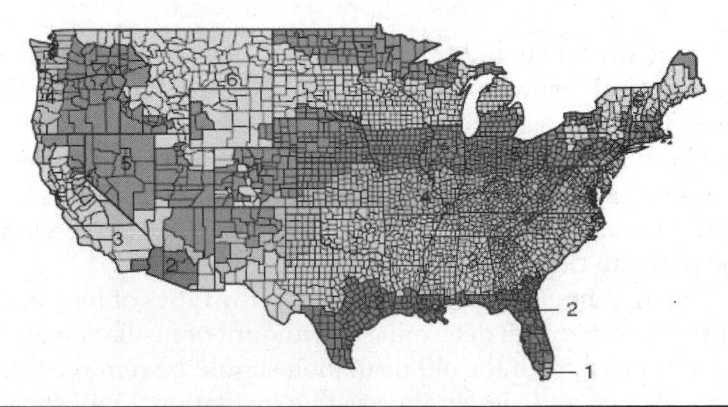

FIGURE 8-14 Insulation map of the contiguous 48 states. *(www1.eere.energy.gov/consumer/tips/insulation.html.)*

All of Alaska is in Zone 7, except for the following boroughs (which are in Zone 8):

- Bethel
- Northwest Arctic
- Dellingham
- Southeast Fairbanks
- Fairbanks North Star
- Wade Hampton
- Nome
- Yukon-Koyukuk
- North Slope

Zone 1 includes:

- Hawaii
- Guam
- Puerto Rico
- Virgin Islands

What is most meaningful to homeowners is familiarity. The fiberglass insulation commonly put into 2 × 4 walls is the familiar R-13 faced insulation (Figure 8-15). R-13 insulation is approximately 3 inches thick with a paper or foil facing. The facing acts as a vapor barrier. FYI: The vapor barrier is installed with the face side toward the hot side of the installation. Any variance to this rule will cause condensation and deterioration of the insulation. A good example of correct installation is R-13 installed in exterior walls with the vapor barrier toward the inside of the home.

FIGURE 8-15 Spray foam insulation. *(http://www.ornl.gov/sci/roofs+walls/insulation/07-G01660.gif.)*

Choosing the Right Type of Insulation for Your Home

Choosing the right type of insulation for your home is imperative. Consider all factors when choosing insulation:

- Climate
- Building design
- Budget
- R values
- Moisture
- Lighting
- Accessibility

In an existing home, consider these factors and then begin in the attic. Adding insulation to an attic is relatively easy and very cost-effective. Most homes require additional insulation. The average attic requires an R value of R-40 to R-60. That is approximately 15 to 22 inches of fiberglass insulation. Remember to use weather stripping on the attic door or access panel.

Insulation is not just to keep heat in the home. Many homeowners in warmer climates do not consider insulation as often as people in cool climates. Insulation is just as important to keep the cool air-conditioned air in your home (Figure 8-16).

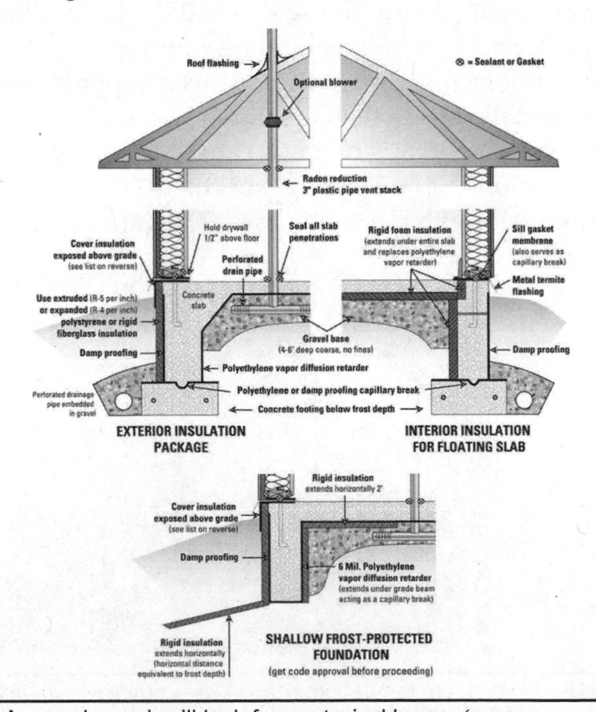

FIGURE 8-16 Areas where air will leak from a typical home. *(www.resourcecenter.pnl.gov.)*

The second space that should be insulated is the basement or crawl space and the home's sill. Once a cap has been put on the home by insulating that attic, stop the second source of air leaking into the home. Different types of insulation can be used together as long as the vapor barrier rule is implemented. The vapor barrier *always* must be on the outside of the insulation toward the heated area (Figure 8-17). Using batts of insulation in conjunction with foam-board insulation is an excellent choice for a crawl space.

FIGURE 8-17 Insulation in an attic.

Unfaced fiberglass batts can be installed between the floor joists, and foam-board insulation (Figure 8-18) can be added on top of the batts to increase the R value. Insulating the attic and basement or crawl space will eliminate the most significant problems of most homes. Interestingly, this problem has been called the *stack effect* (Figure 8-19). The stack effect, dubbed this because the entire house acts like a chimney or stack, can result in a fast-flowing continuous loss of air from the home.

Finally, walls can be insulated. Walls in existing homes can be filled with loose-fill insulation. Small holes are drilled in the walls, and insulation is pumped into the cavities. This will complete the insulation envelope but not the energy envelope.

New home construction is the perfect opportunity to install the optimal amount of insulation. Spray-foam insulation is highly recommended for new home construction (Figure 8-20). Spray-foam insulation can be nontoxic and moisture, bug, and rodent resistant, and it has higher R values than traditional batt or roll insulation. The foam is sprayed into place and expands, leaving no opportunities for air to infiltrate into the home.

FIGURE 8-18 Foam-board insulation.

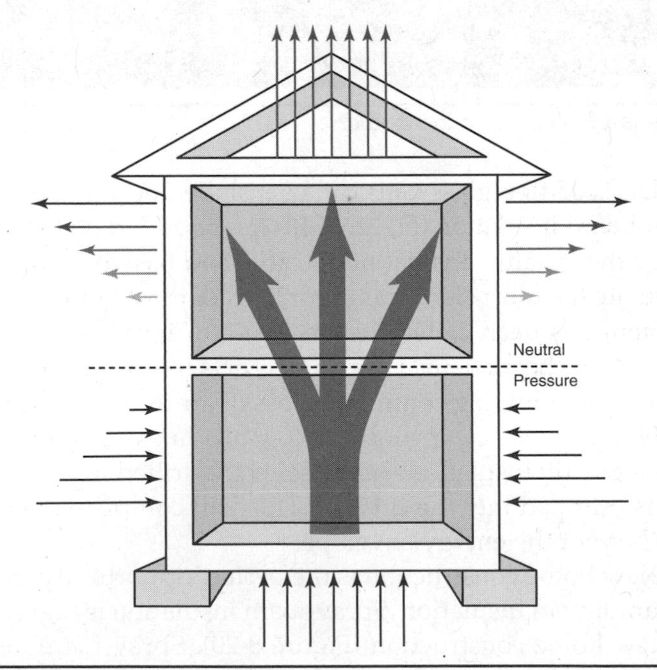

FIGURE 8-19 Stack effect. *(www.hud.gov/offices/pih/programs/ph/phecc/images/stackeff.gif.)*

FIGURE 8-20 Spray-foam insulation. *(www.ornl.gov/sci/roofs+walls/AWT/HotboxTest/WoodFrame/SoyBean/Soy%20Foam%20Wall.jpg.)*

The final insulation option available to homeowners is structural insulated panels (SIPS) (Figure 8-21). The home is designed and built in modules. The modules are shipped to the home site and assembled. Because there are very few penetrations through the exterior walls, there is very little chance of air transfer. SIPS homes are competitively priced, can be built in a few days not months, and are far superior to traditional construction.

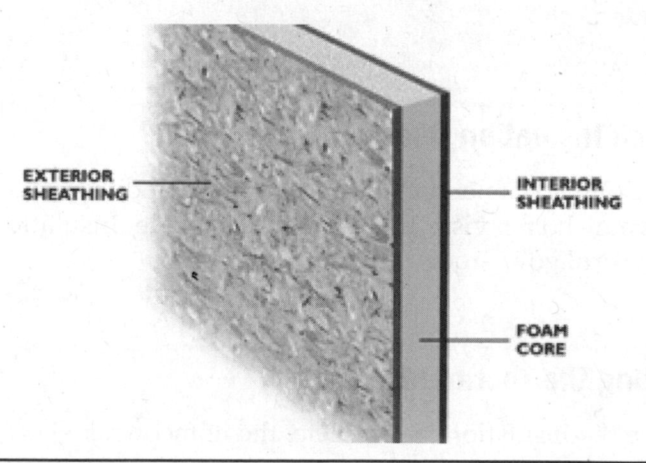

FIGURE 8-21 SIP panel. *(www.energystar.gov/ia/new_homes/behind_the_walls/SIP_panel.jpg.)*

The final choice with almost any type of insulation involves radiant and moisture barriers. These barriers are usually made of materials that will not degrade and will reflect heat back into the home. Foil-faced insulation is a good example of one type of radiant barrier (Figure 8-22).

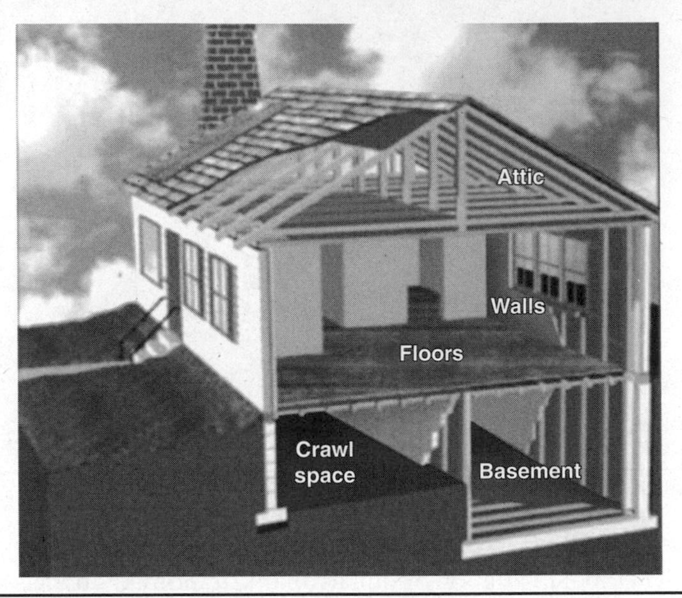

FIGURE 8-22 Foil-faced insulation can be used in all places in a home. *(www1.eere.energy.gov/consumer/tips/insulation_sealing.html.)*

Adding insulation in the areas shown above may be the best way to improve your home's energy efficiency. Insulate either the attic floor or under the roof. Check with a contractor about crawl space or basement insulation.

How Much Insulation Does My Home Need?

For DOE recommendations on how much and where to insulate tailored to your home visit the DOE's Zip Code Insulation Calculator at www.ornl.gov/~roofs/zip/ziphome.html.

Completing the Thermal Barrier

Once the insulation is complete, the homeowner should continue with sealing of the home. The homeowner's goal is to create an energy envelope where the air that enters is controlled by the homeowner (Figure 8-23).

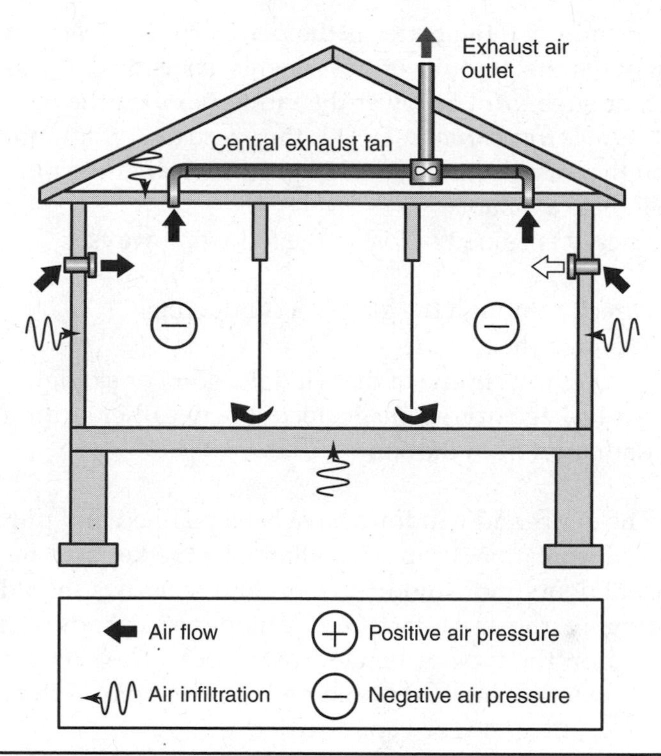

FIGURE 8-23 Proper homeowner-controlled exhaust ventilation. *(www.energysavers.gov/images/exhaust_ventilation.gif.)*

The next step toward a properly sealed house is to address the windows, doors, and skylights. Planning where the windows will be in a home is paramount when designing for energy efficiency. Orientation to the sun depends on the climate. Warmer climates will require the windows to be offset or shaded. Cooler climate will want direct exposure. Planting a tree can solve both problems. A tall tree can block sunlight during the summer months and allow sun into the home during the winter months when the sun is low in the sky.

The homeowner will next want to choose the type of window to be installed. The choice of window is important; the reflection of heat in a cool climate is not what is required. The homeowner cannot rely on the U factor alone but also must check the reflectivity of the windows. The *U factor* is the rate at which a window, door, or skylight conducts nonsolar heat flow, that is, loses heat. The U factor is rated in British thermal units (Btu's) and may refer to the glass, the glazing, or the entire unit. The lower the U factor, the more efficient is the window or door.

Light to solar gain (LSG) is a relative efficiency term that rates the amount of light gained and the amount of heat blocked. The higher the ratings number, the better is the performance. *Solar heat gain coefficient* (SHGC) is the amount of heat energy transmitted through a window, door, or skylight. The lower the rating number, the more efficient is the unit. *Visible transmittance* (VT) is the spectrum of light that is transmitted through a window, door, or skylight. The higher the rating, the more visible light is available.

Energy is gained or lost in the following ways:

1. Directly through the glass via conduction
2. Through the glazing
3. Through the frame of the window, door, or skylight
4. Air leakage from damage, including weather stripping
5. Radiation from the sun

The doors and windows have been planned and purchased; the next step is installation. Proper installation is the key to long-lasting energy-efficient doors and windows. Doors and windows should be installed according to the manufacturers' guidelines. Window installation will depend on the type of home construction. The home may be made of wood, concrete, brick, etc., and each type of home will require an appropriate weather and/or vapor barrier.

Windows

New windows and doors must be installed as an integrated part of the weather barrier. These items must prevent not only air from leaking out but also moisture, wind, and all forms of weather from entering the home (Figure 8-24).

Follow the directions of the window manufacturer as well as the directions from the housewrap manufacturer. *Housewrap* is a directional barrier that allows moisture to evaporate outward while preventing moisture from entering inward. Fold the housewrap into the home, and secure it beyond weather availability. Avoid puncturing or securing the wrap outside the home. Avoid multiple overlapping areas or taping that could cause water to pool.

When the opening is secure and weathertight, coat the surfaces of the window with a caulk sealant. Place the window into the selected area, and tack it in place. Check for level and plumb, and nail or screw the window in place as required by the manufacturer. Add insulation around the window if required. Replace the siding or house treatment to prevent water infiltration. Check the function of the window, and then clean it.

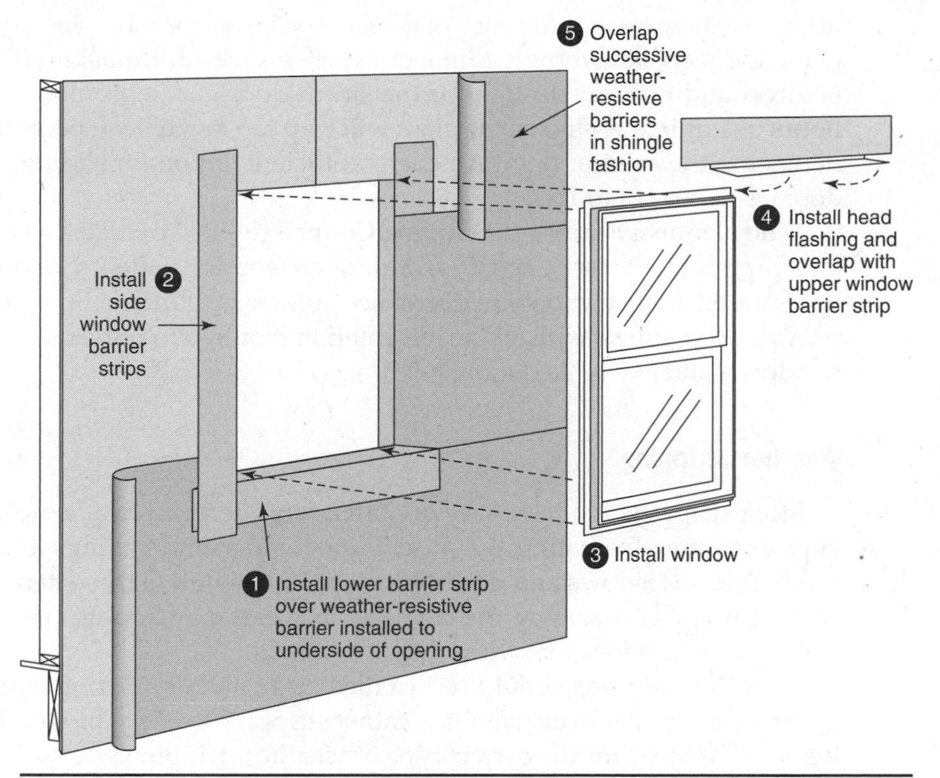

FIGURE 8-24 Flashing around new window installation. *(www.energysavers.gov/your_home/windows_doors_skylights/index.cfm/mytopic=13470.)*

Doors

Exterior door installation is similar to window installation. Selection of the proper door is key. The most common doors today have a steel outside with a foam insulation core. Doors and windows with insulated glass are standard. Insulated glass is two panels of sealed glass with an inert gas such as argon between to prevent moisture and air circulation within the glass. Despite contemporary energy improvements, doors have an R value of only R-5 or R-6. However, this is a major improvement over wood doors, which may have an R value of 1 or 2.

Most doors are *prehung*, which means that the door and frame come as one unit and should be installed as a single unit. Create the space for the new frame and door, and prepare the opening as with a window. Insert the door unit, tack it in place, and check for level. Seal the door with caulk to complete the energy envelope.

One addition that can be made to any doorway is a storm door. This will increase the doorway R factor. Insulated double-pane glass storm doors can increase the efficiency of a doorway. Less expensive single-pane glass doors are not worth the time or expense. One additional limitation is direct and indirect sunlight. If the storm door gets more than a few hours of sunlight each day, the glass will trap too much heat, potentially damaging the exterior door. An energy-efficient option for sliding glass doors is the use of drapes.

The National Fenestration Rating Council (NFRC) operates a voluntary program that tests, certifies, and labels windows, doors, and skylights based on their energy performance ratings. In addition, Energy Star provides consumers with all the information required to purchase doors, windows, and skylights (Figure 8-25).

Weather Stripping

Weather stripping is a necessary but often forgotten part of a home's energy envelope. Insulating the attic, floors, walls, and ceilings and installing new windows and doors are all pointless unless these items are complemented by sealing the home with weather stripping. Weather stripping is very inexpensive and easy to install.

Weather stripping should be installed clean and dry during temperate weather. Similar to insulation, weather stripping needs to be complete. Refer to Table 8-2 for the correct type of weather stripping.

Use the correct type of weather stripping for the application. Most consumers are unaware of the variety of weather stripping available. Verify that the weather stripping is functioning correctly after installation. While weather stripping may seem small and insignificant, without it, all your energy improvements could be rendered ineffective.

1. Caulk and weather-strip doors and windows.
2. Caulk and seal air leaks around plumbing, ducting, or electrical wiring.
3. Caulk penetrations through walls, floors, and ceilings.
4. Install foam gaskets behind outlet and switch plates on walls.
5. Use foam sealant around larger gaps around windows and baseboards.
6. Kitchen exhaust fan covers can keep air from leaking in when the exhaust fan is not in use.

The Roof

Finally, I will address an issue that most homeowners will not. The roof is an area that does not receive any attention until it fails. New products and technology allow for a roof to be very energy efficient (Figure 8-26).

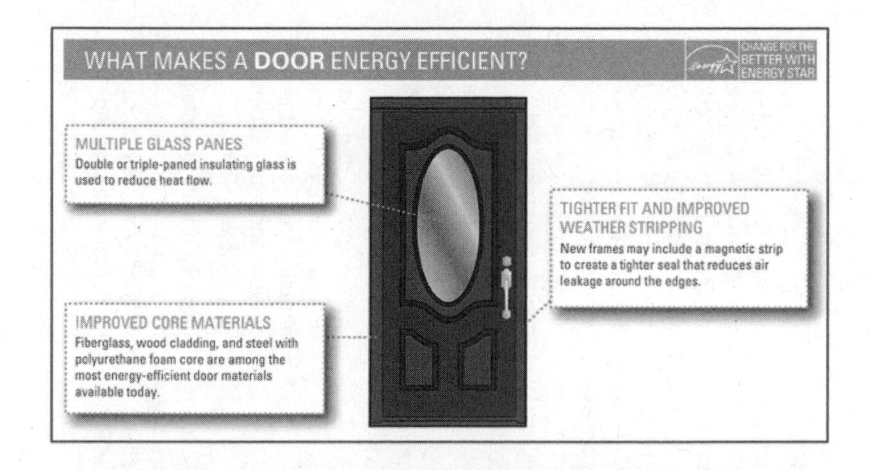

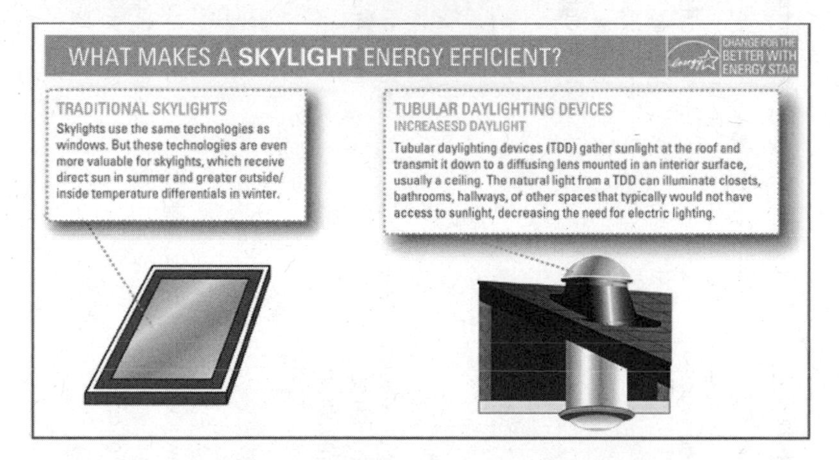

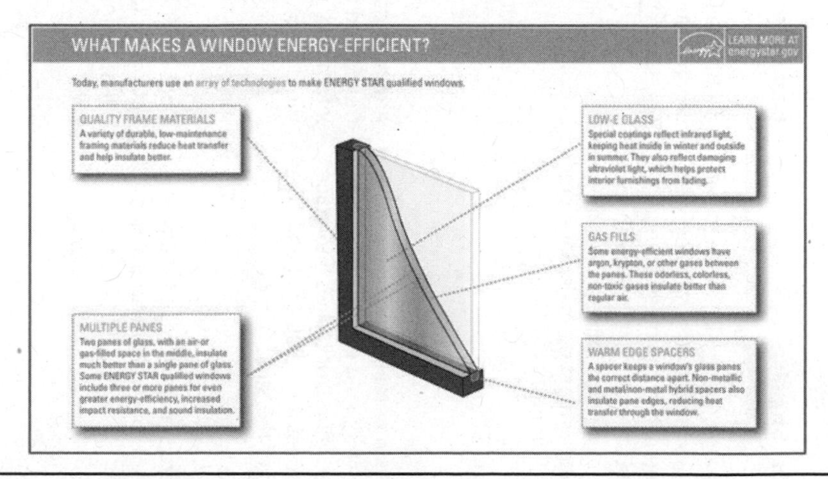

FIGURE 8-25 Energy-efficient door, skylights, and windows. *(www.energystar.gov/index.cfm?c=windows_ doors.pr_anat_window.)*

Table 8-2 Common Weather Stripping

Weather Stripping	Best Uses	Cost	Advantages	Disadvantages
Tension seal: Self-stick plastic (vinyl) folded along length in a V shape or a springy bronze strip (also copper, aluminum, and stainless steel) shaped to bridge a gap. The shape of the material creates a seal by pressing against the sides of a crack to block drafts.	Inside the track of a double-hung or sliding window, top and sides of door.	Moderate; varies with material used.	Durable. Invisible when in place. Very effective. Vinyl is fairly easy to install. Look of bronze. Works well for older homes.	Surfaces must be flat and smooth for vinyl. Can be difficult to install because corners must be snug. Bronze must be nailed in place (every 3 inches or so) so as not to bend or wrinkle. Can increase resistance in opening/closing doors or windows. Self-adhesive vinyl available. Some manufacturers include extra strip for door striker plate.
Felt: Plain or reinforced with a flexible metal strip; sold in rolls. Must be stapled, glued, or tacked into place. Seals best if staples are parallel to length of the strip.	Around a door or window (reinforced felt); fitted into a door jamb so that the door presses against it.	Low.	Easy to install, inexpensive.	Low durability; least effective preventing airflow. Do not use where exposed to moisture or where there is friction or abrasion. All-wool felt is more durable and more expensive. Very visible.
Reinforced foam: Closed-cell foam attached to wood or metal strips.	Door or window stops; bottom or top of window sash; bottom of door.	Moderately low.	Closed-cell foam an effective sealer; scored well in wind tests. Rigid.	Can be difficult to install; must be sawed, nailed, and painted. Very visible. Manufacturing process produces greenhouse gas emissions.

Weather Stripping	Best Uses	Cost	Advantages	Disadvantages
Tape: Nonporous, closed-cell foam, open-cell foam, or ethylene propylene diene monomer (EDPM) rubber.	Top and bottom of window sashes, door frames, attic hatches, and inoperable windows. Good for blocking corners and irregular cracks.	Low.	Extremely easy to install. Works well when compressed. Inexpensive. Can be reinforced with staples.	Durability varies with material used, but not especially high for all; use where little wear is expected; visible.
Rolled or reinforced vinyl: Pliable or rigid strip gasket (attached to wood or metal strips).	Door or window stops; top or bottom of window sash; bottom of a door (rigid strip only).	Low to moderate.	Easy installation. Low to moderate cost. Self-adhesive on pliable vinyl may not adhere to metal; some types of rigid strip gaskets provide slot holes to adjust height, increasing durability. Comes in varying colors to help with visibility.	Visible.
Door sweep: Aluminum or stainless steel with brush of plastic, vinyl, sponge, or felt.	Bottom of interior side of in-swinging door; bottom of exterior side of exterior-swinging door.	Moderate to high.	Relatively easy to install; many types are adjustable for uneven threshold. Automatically retracting sweeps also available, which reduce drag on carpet and increase durability.	Visible. Can drag on carpet. Automatic sweeps are more expensive and can require a small pause once door is unlatched before retracting.

(continued on next page)

Table 8-2 Common Weather Stripping (continued)

Weather Stripping	Best Uses	Cost	Advantages	Disadvantages
Magnetic: Works similarly to refrigerator gaskets.	Top and sides of doors, double-hung and sliding window channels.	High.	Very effective air sealer.	
Tubular rubber and vinyl: Vinyl or sponge rubber tubes with a flange along length to staple or tack into place. Door or window presses against them to form a seal.	Around a door.	Moderate to high.	Effective air barrier.	Self-stick versions challenging to install.
Reinforced silicone: Tubular gasket attached to a metal strip that resembles reinforced tubular vinyl.	On a doorjamb or a window stop.	Moderate to high.	Seals well.	Installation can be tricky. Hacksaw required to cut metal; butting corners pose a challenge.
Door shoe: Aluminum face attachment with vinyl C-shaped insert to protect under the door.	To seal space beneath door.	Moderate to high.	On the exterior, product sheds rain. Durable. Can be used with uneven opening. Some door shoes have replaceable vinyl inserts.	Fairly expensive; installation moderately difficult. Door bottom planning possibly required.
Bulb threshold: Vinyl and aluminum.	Door thresholds.	Moderate to high.	Combination threshold and weather-strip; available in different heights.	Wears from foot traffic; relatively expensive.

Weather Stripping	Best Uses	Cost	Advantages	Disadvantages
"Frost-brake" threshold: Aluminum or other metal on exterior, wood on interior, with door bottom seam and vinyl threshold replacement.	To seal beneath a door.	Moderate to high.	The use of different materials means less cold transfer. Effective.	Moderately difficult to install, involves threshold replacement.
Fin seal: Pile weather-strip with plastic Mylar fin centered in pile.	For aluminum sliding windows and sliding glass doors.	Moderate to high.	Very durable.	Can be difficult to install.
Interlocking metal channels: Enables sashes to engage one another when closed.	Around door perimeters.	High.	Exceptional weather seal.	Very difficult to install as alignment is critical. To be installed by a professional only.

Source: www.energystar.gov/index.cfm?c=home_sealing.hm_improvement_sealing.

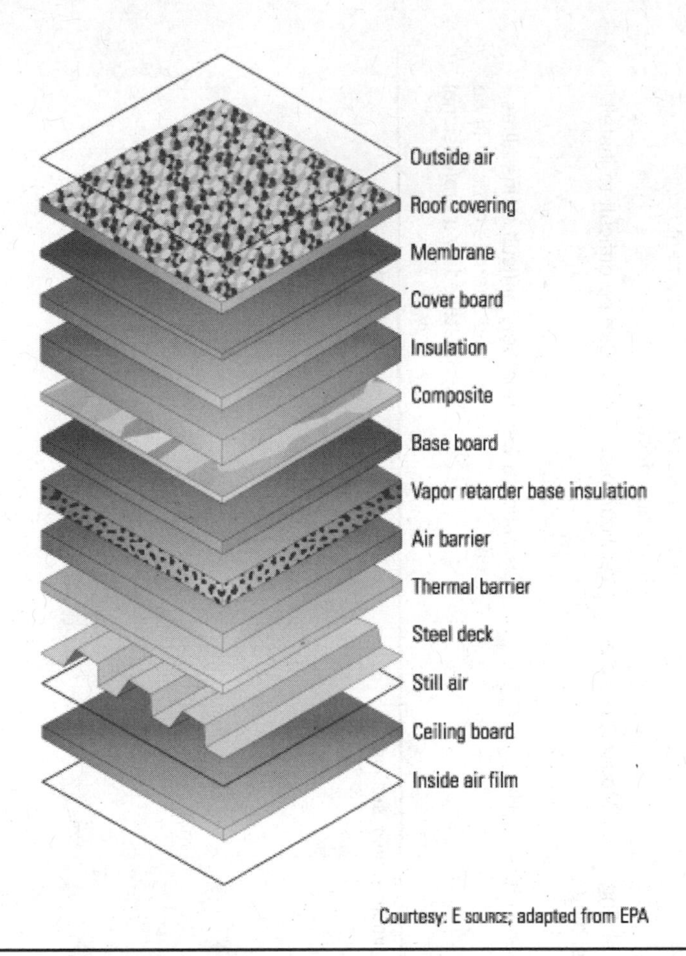

FIGURE 8-26 Typical exterior wall home construction. *(www.energystar.gov/index.cfm?c=windows_doors.pr_windows.)*

Before installing the same old products on your home, check the Energy Star Web site for energy-efficient possibilities.

This brings us full circle—the original science lesson (Figure 8-27). Once the weather stripping is complete, you should have a home that looks and acts like Figure 8-28 on page 184.

This home has a complete energy envelope. Ventilation and air exchange are controlled. Heating and cooling losses are limited to the amount of fresh air required in the home. This type of home is a home that every person can own today.

Next, I will define some of the terms that I have used and outline a few more free or low-cost home improvements. I also reference the Web to allow you to expand your knowledge. Knowledge is what you require; the "doing" is cost-effective and easy.

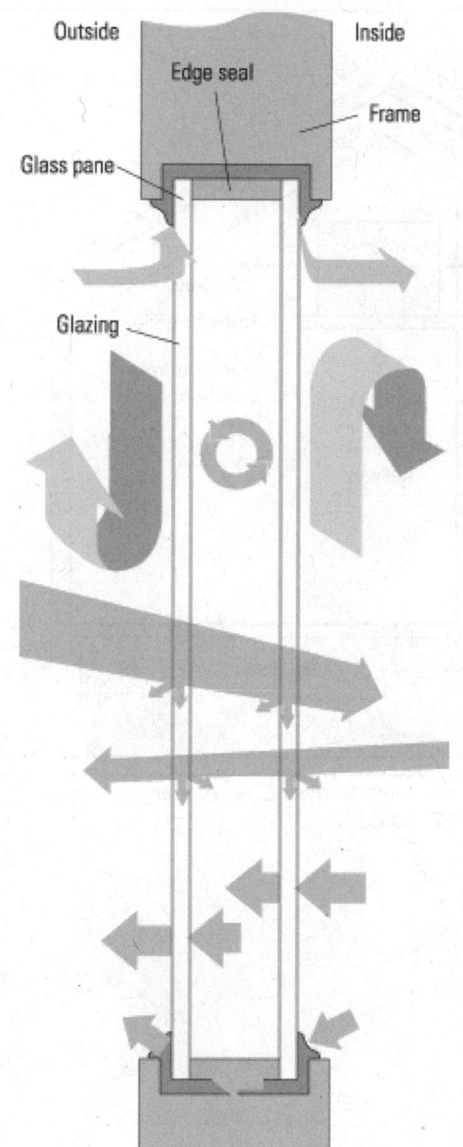

Infiltration
Air leaks around the frame, around the sash, and through gaps in movable window parts. Infiltration is foiled by careful design and installation (especially for operable windows), weather stripping, and caulking.

Convection
Convection takes place in gas. Pockets of high-temperature, low-density gas rise, setting up a circular movement pattern. Convection occurs within multiple-layer windows and on either side of the window. Optimally spacing gas-filled gaps minimizes combined conduction and convection.

Radiation
Radiation is energy that passes directly through air from a warmer surface to a cooler one. Radiation is controlled with low-emissivity films or coatings.

Conduction
Conduction occurs as adjacent molecules of gases or solids pass thermal energy between them. Conduction is minimized by adding layers to trap air spaces, and putting low-conductivity gases in those spaces. Frame conduction is reduced by using low-conductivity materials such as vinyl and fiberglass.

Courtesy: E SOURCE

FIGURE 8-27 Science again: conduction, radiation, convection, and infiltration. *(www.energystar.gov/ia/business/Web_art/EPA-BUM-SupLoads_7-2.gif.)*

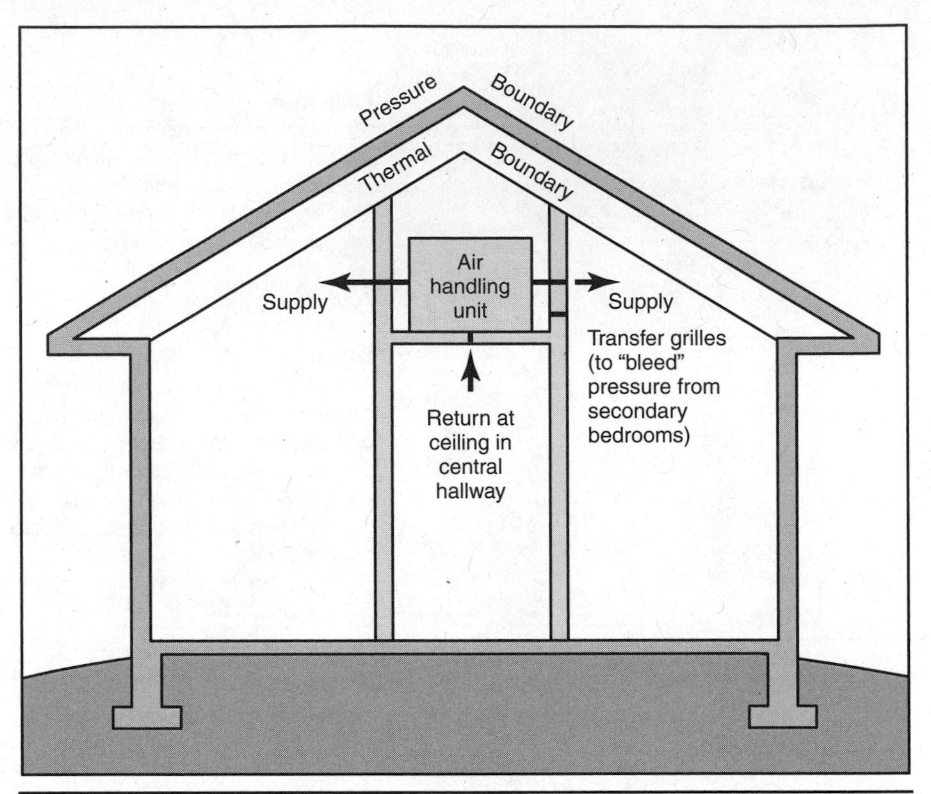

FIGURE 8-28 Equalizing air pressure efficiently. *(www1.eere.energy.gov/buildings/residential/images/hvac_3.jpg.)*

CHAPTER 9

Standards, Conversions, and Green References

Is there any dispute that wealthy people live more comfortably than less monetized people? The reason is not money, but information. Are wealthy people more informed than less affluent individuals? Maybe, but probably not. Wealthy people can hire the expertise they require to allow themselves to live more comfortably.

I cannot make you wealthy, but I can inform you so that you can make intelligent choices about your home and your purchases. I also will not attempt to make you a science major or a Nobel Prize winner, but I do need to review some simple scientific information with you.

Most people believe that they understand electricity, their utility bills, and the Energy Star rating. While most individuals have some understanding of this information, their knowledge is incorrect or incomplete. I will provide you with the basics to allow you to make informed decisions.

This chapter will review the common terms and standards used in this book. Simple, clear definitions will be presented. Most of the following terms are more complex, so additional references will be supplied. For the use of most homeowners, the definitions provided will be adequate.

Temperature and Pressure

Relative humidity Relative humidity is the amount of water vapor that exists in the air at a known temperature. The amount of water contained by the air is relative to the temperature. Relative humidity is somewhat more complex, but this definition will be sufficient for the average homeowner.

Dew point The dew point is related to relative humidity. The dew point is the point at which water will condense from the air at a constant barometric pressure. This is also known as the *saturation point*. The dew point is related to relative humidity. A higher relative humidity indicates that the air contains more water than at a lower relative humidity. A relative humidity reading of 100 percent indicates that the dew point is equal to the current temperature. When the dew point falls below freezing, the common name is the *frost point*.

Barometric pressure Barometric pressure is a measure of atmospheric pressure. Barometric pressure is the force exerted against an object at a certain level above sea level. The higher you travel in the atmosphere, the lower is the atmospheric pressure. This is why airplanes need to be pressurized. Barometric pressure is commonly measured as 29.92 inches of mercury (inHg), 14.696 pounds per square inch (psi), or 1013.25 millibars (mbar) at sea level. These readings should be familiar to you from your local weather forecast. Barometric pressure is important to homeowners because the pressure differential (the pressure difference between the inside and the outside of the home) is one of the factors that causes air to flow from a home, therefore reducing the home's efficiency.

Energy and Units

The British thermal unit (Btu) is defined as the amount of heat required to raise the temperature of 1 pound of liquid water by 1°F at a constant pressure of 1 atm. More commonly, the number of British thermal units is the amount of energy in a product. For example, a gallon of gasoline has approximately 125,000 Btu's.

The British thermal unit is a unit of energy, used to describe the heat value of a substance. Thus:

1 watt is approximately 3.413 Btu/h.
1,000 BTU/h is approximately 293 watts.

One short ton of cooling in North American refrigeration and air-conditioning applications is 12,000 BTU/h. This is the amount of energy required to melt one short ton of ice in 24 hours. Converted to watts, this is approximately 3.51 kilowatts.

What Is a Watt (W)?

A *watt*, named after James Watt (1736–1819), is a unit of power equal to 1 joule per second. It is also the power dissipated by a current of 1 ampere flowing across a resistance of 1 ohm. The concept of a watt is very meaningful to most people, but not in this form. Most people relate to the energy of a watt best through their appliances. Most compact fluorescent lights (CFLs) are 10 to 30 watts, most incandescent bulbs are 40 to 100 watts, and your hair dryer is 1,000 to 1,200 watts. I will use a 100-watt incandescent light bulb as the standard.

What Is a Kilowatt (kW)?

A *kilowatt* is 1,000 watts. If you turn on your 100-watt bulb, you are using 100 watts, or 1/10 of a kilowatt.

What Is a Kilowatthour (kWh)?

A *kilowatthour* is 1,000 watts used for 1 hour of time. For example, if you boil water using 1,000 watts for 1 hour, you have used 1 kilowatt of power. If you turn on your 100-watt light bulb for 10 hours, then you will have used 1 kilowatt.

The Therm

The *therm* is a unit of energy in the United States and the European Union. One therm equals 100,000 British thermal units. The United States uses the standard British thermal unit scale, and the European Union uses the international scale of British thermal unit energy.

According to Internet encyclopedia *Wikipedia*:

- In natural gas, by convention, 1 MMBtu (1 million British thermal units, sometimes written *mmBtu*) = 1.054615 gigajoule (GJ). Conversely, 1 gigajoule is equivalent to 26.8 m^3 of natural gas at defined temperature and pressure. Thus 1 MMBtu = 28.263682 m^3 of natural gas at defined temperature and pressure.
- One standard cubic foot of natural gas yields ≈ 1030 British thermal units (between 1010 and 1070 British thermal units, depending on quality when burned).

The *therm* (symbol *thm*) is a non-SI unit of heat energy equal to 100,000 British thermal units (Btus). It is approximately the energy from burning 100 cubic feet (often referred to as 1 Ccf) of natural gas.

- Therm (EC) ≈ 100,000 Btu_{IT}
  = 105,506,000 joules
  ≈ 29.3072222 kilowatthours

  The therm (EC) is often used by engineers in the United States.
- Therm (US) ≈ 100,000 $BTU_{59°F}$
  = 105,480,400 joules
  ≈ 29.3001111 kilowatthours
- Therm (UK) = 105,505,585.257348 joules
  ≈ 29.30710701583 kilowatthours

U Factor

The U factor is the rate at which a window, door, or skylight conducts nonsolar heat flow, that is, loss of heat. The U factor is rated in British thermal units and may refer to the glass, the glazing, or the entire unit. The lower the U factor, the more efficient is the window or door.

Cubic feet per minute (CFM) is used to measure the amount of air movement.

Solar energy has a few unique terms: *Light to solar gain* (LSG) is a relative efficiency term that rates the amount of light gained and the amount of heat blocked. The higher the rating number, the better the performance. *Solar heat gain coefficient* (SHGC) is the amount of heat energy transmitted through a window or door. The lower the rating, the more efficient is the unit. *Visible transmittance* (VT) is the spectrum of light that is transmitted through a window. The higher the rating, the more visible light is available.

R Value

The *R value* is the thermal resistance to heat. The higher the number, the better is the resistance.

Ton: Short, Metric, and Long

A *short ton* is 2,000 pounds (907 kilograms), and this unit is used commonly in the United States. A *metric ton* is approximately 2,204 pounds (1,000 kilograms). A *long ton* is approximately 2,240 pounds (1,016 kilograms).

Now let's take a short break and accomplish some immediate cost savings.

General Items

Change Your Driving Habits

Do not accelerate rapidly; this wastes fuel for no reason. The fastest way to travel is with consistency. Remember the old tortoise and hare story? The more consistent your speed and the fewer number of times that you slow down or come to a stop, the faster you will reach your destination. This includes using your brakes. Often drivers brake unnecessarily on the highway. Every time you step on your brake, you are losing money. Pay attention when you drive, and you will find that you do not need to apply the brakes as frequently as you currently do.

Use the Correct Fuel in Your Car

Did you know there are three types of gas and diesel fuel? "Why?" is the more important question. For this, you must read the manufacturer's manual for your car. This manual will indicate the correct type of gas for your car. There are three grades of gasoline: regular, premium, and ultra. All three grades formerly contained lead. What is important in selecting a grade of gasoline is the following:

- Potential energy
- Level of refinement
- Octane rating
- Cost

Regular gas has the most potential energy, has been refined the least, has the lowest octane rating, and is the lowest in price. Premium gas has the second most energy, is refined a little more than regular gasoline, and has a slightly higher octane rating and cost. Ultra gas has the least potential energy, is refined the most, and has the highest octane rating and cost.

Your car is designed to use a specific octane rating. If you do not use the correctly rated gasoline, the mileage and performance of your car may suffer severely. If you use premium fuel in your car and the car requires regular gasoline, you actually may be spending extra money and causing poor performance. If you are putting regular fuel into a car that requires premium fuel, you will cause problems and actually could cause engine damage. The automobile is design for an appropriate fuel. Always use the fuel recommended by the manufacturer.

Buy Local

Buy local. This is very important. It is unnecessary to purchase broccoli from California if you live in New York. Less energy can be expended to obtain the same or better-quality products, especially food. If tomatoes are shipped from California, this means that the food has been picked before it was ripe, allowing the shipper time to bring this item to market. This is why many fruits and vegetables are tasteless.

Buy Energy Star

As you should realize by now, Energy Star products will save you more money over the life of the product when compared to non–Energy Star products. In many instances, the energy saved with an Energy Star appliance will actually pay for the appliance over its useful life.

The current Energy Star rating system involves many factors. It is based on statistical averages and assumptions. What this means to the consumer is that the rating system is not perfect and not correct for everyone. If you need unique products or unusual services, it is best to perform your own calculations. For most consumers, however, the Energy Star system is excellent.

Refrigerators, washers, dryers, air conditioners, and so on all use use electricity in various way and in variable degrees. The more you open your refrigerator, the more often it will run to cool your food. The best that can be done is to figure the maximum energy the appliance uses, compare that with each product you prefer, and factor in how you will use the appliance.

The current Energy Star label, EnergyGuide, is good. Always buy the most cost-effective product in the category; it will cost you less over the lifetime of the product. Understanding your requirements as a consumer will allow for correct product selection. The federal government is developing a new set of Energy Star ratings that promise to be more effective for all consumers.

If you need to know about the rating system of a product, go to the Energy Star Web site and check the product and its ratings: www.energystar.gov/index.cfm. Research and develop your own Energy Star number that will allow for the purchase of a computer that meets your technical and energy requirements.

What exactly is the Energy Star rating program? According to Energy Star: "Energy Star is a joint program of the U.S. Environmental Protection Agency and the U.S. Department of Energy helping us all save money and protect the environment through energy-efficient products and practices."

In 1992, the U.S. Environmental Protection Agency (EPA) introduced Energy Star as a voluntary labeling program designed to identify and promote energy-efficient products to reduce greenhouse gas emissions. Computers and monitors were the first labeled products. The Energy Star label is now on over 50 product categories, including major appliances, office equipment, lighting, and home electronics. The EPA also has extended the label to cover new homes and commercial and industrial buildings.

A typical household spends $2,000 a year on energy bills. With Energy Star, you can save more than 30 percent, or more than $700 per year, with similar savings of greenhouse gas emissions, without sacrificing features, style, or comfort. Energy Star helps you to make the energy-efficient choice.

If you are looking for new household products, look for ones that have earned the Energy Star label. They meet strict energy-efficiency guidelines set by the EPA and Department of Energy (DOE). You can identify them by the blue Energy Star label (Figure 9-1).

FIGURE 9-1 Energy Star label. *(www.Energystar.gov.)*

If you are looking to buy a new home, look for one that has earned the Energy Star rating.

If you are looking to make large improvements to your home, the EPA offers tools and resources to help you plan and undertake projects to reduce your energy bills and improve home comfort.

The EPA's Energy Star partnership for businesses offers a proven energy management strategy that helps in measuring current energy performance, setting goals, tracking savings, and rewarding improvements. The EPA also provides an innovative energy performance rating system to see how your energy use compares with that of similar buildings and plants. The EPA also recognizes top-performing buildings with the Energy Star rating.

Most consumers are familiar with the Energy Star label but not so familiar with the way the rating system works. Figure 9-1 shows the most common Energy Star label, which is used to identify Energy Star–certified products. Some variations on labeling are as follows:

- Figure 9-2 shows the retail display commonly used in "big box" stores. This is another variation of the Energy Star label that will alert consumers to Energy Star products.
- The Energy Star Tax Credit Label (Figure 9-3) identifies when tax credits or rebates apply to a purchase.
- The final label, the EnergyGuide label (Figure 9-4), provides the consumer with information about the specific product. The information contained will help the consumer to choose the most energy-efficient purchase. Most Energy Star products will pay for themselves, and save you money, over the life of the appliance. The distinctive yellow and black EnergyGuide labels appear on Energy Star appliances. These products are subject to minimum efficiency standards set by the federal government. Consumers will find these labels only on appliances that meet the Energy Star requirements.

FIGURE 9-2 Energy Star label. *(www.energystar.gov.)*

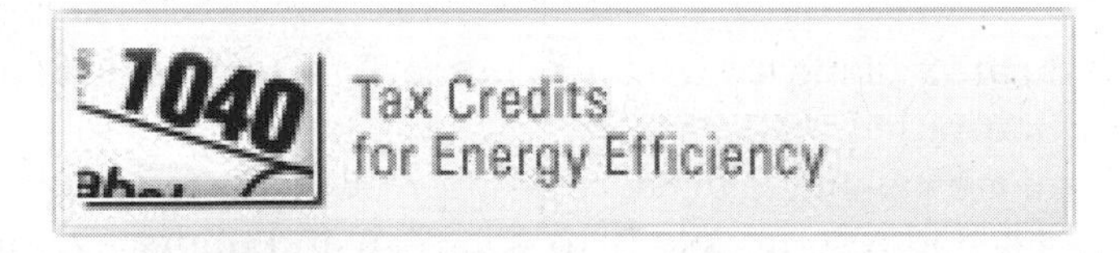

FIGURE 9-3 Energy Star Tax Credit label. *(www.energystar.gov.)*

Current Energy Star ratings exempt some products. Televisions, clothes dryers, ranges and ovens, and space heaters have to meet federal minimum efficiency standards; they were exempted from the EnergyGuide program. The amount of energy these products use does not vary substantially from model to model.

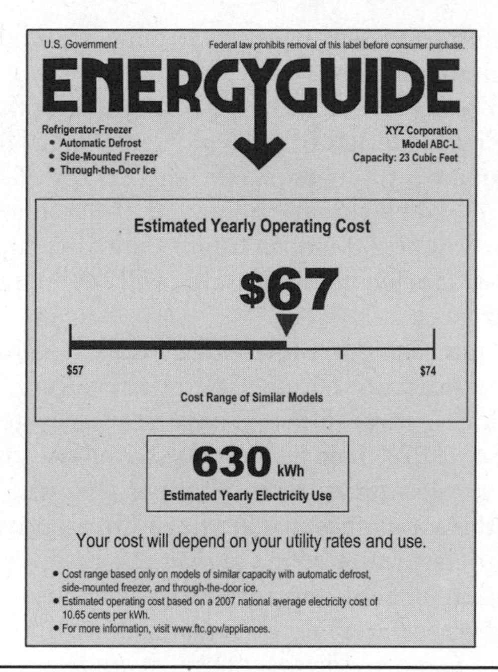

FIGURE 9-4 EnergyGuide label. *(www.ftc.gov/opa/2007/08/elabel.jpg.)*

Reading the EnergyGuide label is simple. In short, the smaller the number, the more you will save during operation of the appliance. Beyond this oversimplified explanation, EnergyGuide labels come in slightly different formats for different appliances. The information contained within the label will help you to choose the most efficient appliance for your use.

The left side of the label, under the header "EnergyGuide," is a description of the appliance. The appliance whose label is in the figure is a refrigerator-freezer. The right side lists the manufacturer, the model number, capacity, and other general product information. This unit has a capacity of 23 cubic feet. The first text box shows the annual operating cost on average. This model will cost $67 per year to operate. The second text box estimates the amount of energy this model refrigerator will use annually. This model will use 630 kilowatt hours of electricity.

This is the key to the labeling system. Consumers pay different rates for electricity. The annual operating cost is only an average. Know how much you pay per kilowatt, and you can calculate the actual operating cost based on average use in your area. Energy Star uses a range of similar models, average use, and an average cost of electricity to determine these estimates. Knowing your personal habits, electrical rates, and requirements allows you to determine the best choice for your use. The labeling system permits consumers to choose among models that are all rated in a similar fashion.

One difficulty with the Energy Star system is that the labels are never updated. Newer models of appliances may seem to fall within the same category but are actually more efficient. Energy Star also chooses a range of appliances, so be sure to compare a 23-cubic-foot refrigerator with a unit of the same size. Manufacturers sometimes upgrade or downgrade their original efficiency ratings. This will not be reflected in the current Energy Star label.

When purchasing appliances, be aware of other rating systems and information. In addition to the Energy Star rating, be aware of the product's functions. Air conditioners have a rating system called the *energy efficiency rating* (EER). This would be a secondary source of information to assist you in selecting an air conditioner. Washing machines specifically are one of the appliances that are not addressed well by the Energy Star system. Understanding your usage in this case would be a better source of information for choosing a washing machine.

Consumer beware! Energy Star must rate the water and energy used for a washer and provide a statistical average of savings. This is another flaw in the Energy Star system. If you wash your clothing in hot water only, the Energy Star rating will be completely useless because it is based on averages.

I suggest using the Energy Star label to compare similar-sized models. Then take into consideration how you personally will use the appliance. Check any additional information or rating systems and choose the appliance that best fits your requirements. CBS News has an excellent interview and discussion about the few anomalies in the Energy Star system. This can be viewed at www.cbsnews.com/video/watch/?id=1455258n. Energy Star labeling is excellent for most consumers, but being aware of your own personal usage patterns is just as important.

Tax credits and rebates are another benefit to buying Energy Star–certified products. Energy Star products that have added tax benefits are identified with the Energy Star Tax Credit label (see Figure 9-3). If the label is absent from the product you are looking at, ask the sales representative or go to the Energy Star Web site to check for rebates.

In addition to products, services may be entitled to tax credits or rebates. Products that require specialty installation may have tax credits available with them. Such items include:

- Water heaters
- Central air-conditioning systems
- Heat pumps

- Solar panels
- Wind energy devices
- Heating systems
- Fuel cells
- Wood or pellet stoves

Some of these products do have limitations for installation. For example, the first $2,000 of the cost of a heating system may be tax deductible. If the cost of the installation is more than that, there is no rebate or credit over the first $2,000. The rebate figures are based on installation averages. If the contractor wishes to charge multiples over the tax rate, get more quotes; the first contractor may be charging exorbitant fees.

The second item to confirm is the contractor's credentials. The installation professional is required to be Energy Star–certified. If the professional is not certified, then the homeowner cannot receive a rebate or tax credit for the installation. Check with the remodeler and with the Energy Star Web site.

A list of items that cannot receive tax credits for installation also can be found on the Energy Star Web site. These items include:

- Insulation
- Doors
- Windows
- Roofs

A tax credit may not be available for installation of these items, but there may be tax credits for the purchase of the products or materials themselves.

Energy Star is creating new standards for energy efficiency and a more effective rating system. The new standards and developments can be found at www.energystar.gov/index.cfm?c=new_specs.new_prod_specs. The current standards are available at www.energystar.gov/index.cfm?c=prod_development.prod_development_index. While Energy Star is a good rating system, it cannot be perfect for all appliances and situations. Know your personal requirements, and remember, a low number almost always means a better, more efficient product and a better value for you.

The EER is the Air-Conditioning and Refrigeration Institute standardized rating. This reports central air-conditioning efficiency at 80°F indoors and 95°F outdoors. This rating measures steady-state efficiency. The higher the EER number, the more effective is the product.

The Seasonal Energy Efficiency Rating (SEER) is available for residential central air conditioners and generally is considered a more reliable indicator of the overall energy efficiency than the EER. The higher the SEER number, the more effective is the product.

Global Warming

Green Anything related to good or clean energy products or solutions.
Climate change Used to refer to dire changes in the environment.
Global warming The heating of the earth owing to human causes.
Carbon cycle The interrelation of carbon as it is used in the macro environment.
PPM (also *ppm*) Parts per million.
PPT (also *ppt*) Parts per thousand.

Rather than bore you with thousands of additional terms, from the sunny state of California comes a great Web site for energy terminology. This information can be found at www.consumerenergycenter.org/glossary/e.html#EER_(energy_efficiency_ratio). In addition, do not forget my favorite and most reliable Web site, www.energystar.gov, and the equally valuable www.DOE.gov and www.EPA.gov sites.

I have found all these Web sites to be credible and up-to-date. With technology changing so rapidly, it is important to have both up-to-date information and reliable and credible information. I have made every attempt to validate all the Web sites and resources referred to in this book. If you have any additional links or reference information that you would like to see in future printings, please send me a note at HEAinfo@exploresynergy.org.

Additional resources for my next subject, funding your home improvements, can be found at Synergy (www.exploresynergy.org). Funding is usually the factor that determines if a project will be attempted. In my final chapter, I will review the many options available to all types of homeowners.

CHAPTER 10

Funding Your Green Home-Improvement Projects

How you fund your home improvements and green-technology projects is completely your choice. I will attempt to provide you with the optimal ways to finance your home remodel. Many governments offer funding in various forms for "green energy." Green energy includes conservation and items such as insulation and compact fluorescent lights (CFLs). This is probably the best time in history to remodel and use energy-efficiency services and products in your home.

In this chapter, I will demonstrate the financial options available to you (Figure 10-1) and provide you with the sound reasons for each option.

Let's take a short break and accomplish some immediate cost savings.

Analyze Your Lighting

This is our final no-cost home-improvement project. Find a pen, paper, and a child, if you have one. Then go from room to room in your house and make a list all the lights that you have in your home (Figure 10-2). You will want to write down the wattage of each bulb. An average home uses 15 to 20 percent of its electricity for lighting. You may be able to reduce this amount by 75 to 90 percent. Remember, CFLs last seven times longer than incandescent light bulbs and can be many times more efficient than ordinary bulbs. In addition to the energy and money savings, this equates to less time buying and changing bulbs and less waste going into the landfill. Government and corporate sponsorship of these energy-efficient bulbs means that you usually can receive these bulbs for free or at reduced cost. Look for in-store or mail-in rebate, or use the tax credit or rebate available for buying these bulbs.

Figure 10-1 Funding your green home improvements in the new economy. *(http://kline.house.gov/images/user_images/money_down_drain.jpg.)*

Figure 10-2 Your green assistant. *(www.publicdomain pictures.net.)*

Table 10-1 is the result of this analysis of my own home lighting. You should do the same examination for your home. You probably will have approximately similar results in savings.

Table 10-1 Home Lighting Analysis

	Original		Total			Total	Light
	No. of Bulbs	**Watts**	**Watts**	**No. of Bulbs**	**Watts**	**Watts**	**Description**
Master Bedroom	2	150	300	2	23	46	Table
Entrance	1	75	75	1	11	11	Table
	1	60	60	1	1	1	Table
	1	300	300	1	23	23	Floor

	Original		Total			Total	Light
	No. of Bulbs	Watts	Watts	No. of Bulbs	Watts	Watts	Description
Living room	2	100	200	2	14	28	Floor
	4	100	400	4	14	56	Ceiling
	1	60	60	1	14	14	Table
	1	40	40	1	7	7	Table
Stairwell	3	100	300	1	14	14	Ceiling
Kitchen	5	120	600	5	23	115	Recessed
	1	40	40	1	7	7	Stove
	2	60	120	2	11	22	Counter
	1	7	7	1	1	1	Night
Bath 1	4	40	160	4	11	44	Vanity
	1	100	100	1	14	14	Ceiling
Bedroom 2	1	300	300	1	23	23	Floor
	1	100	100	1	23	23	Table
Bedroom 3	3	120	360	3	23	69	Recessed
	1	40	40	1	7	7	Table
Attic	1	100	100	1	14	14	Table
Basement	4	120	480	4	23	92	Recessed
	4	40	160	4	11	44	Table
Bedroom 4	4	120	480	4	23	92	Recessed
	1	60	60	1	14	14	Table
Theater room	5	120	600	5	23	115	Recessed
	1	60	60	1	11	11	Table
	1	7	7	1	1	1	Night
Bath 2	1	100	100	1	14	14	Ceiling
	2	75	150	2	11	22	Vanity
Outside light	3	100	300	3	14	42	Front
	1	100	100	1	14	14	Front
	8	120	960	8	23	184	Spot
	1	100	100	1	14	14	Side door
	1	100	100	1	14	14	Side door
	1	100	100	1	14	14	Back door
	1	100	100	1	14	14	Decorative
Garage	2	60	120	2	11	22	Garage door
	3	100	300	3	14	42	Overhead
	2	100	200	2	14	28	Workbench
Totals	83	3,694	8,139		554	1,332	

Energy savings 84%

Money savings each year $226.80

Per life of the bulb $1,567.60

Funding Your Project

Your first financial consideration when doing a home remodel is your savings. The money that you have accumulated can be put to work to make your dreams come true (Figure 10-3).

FIGURE 10-3 Better ways to use your money. *(www.publicdomainpictures.net.)*

For many of the projects in this book, you will have noticed, spending a little money now can reap large financial rewards in the future. Some of the options available to you, for example, solar energy, actually could produce a monthly income.

Using your savings is the friendliest way to accomplish your project. There are no loan applications, interest rates, or paperwork. In addition to these advantage, if you pay in cash, many contractors or suppliers will offer you a better deal. If you are spending your own hard-earned after-tax income, be sure to check for rebates or tax incentives from the government. In the United States, more information can be found at:

The Department of Energy (DOE): www.energy.gov/taxbreaks.htm
Energy Star Program: www.energystar.gov/index.cfm?fuseaction=rebate.rebate_locator
Internal Revenue Service: www.irs.gov/

You will find financial incentives to replace your furnace, central air conditioner, all your appliances, and your car. You name it, and you can find financial benefits for almost all energy items in your home. Be sure to check with your state and local governments as well. You often will find local programs that assist consumers.

If you are spending your own money for your remodel or energy-efficient purchase, be sure to ask about the details from the sales or service provider.

Funding Options

The most common options available to consumers are traditional loans. Traditional loans come in many forms:

1. Friends or family
2. Conventional
3. Low interest
4. Zero interest
5. Deferred
6. Home equity
7. Remortgage

Each type of loan has its advantages and disadvantages. The first type of loan is from your friends or family. This loan is easy to obtain, and all the details of the loan are negotiable. One quandary is that you may not want friends or relatives to be familiar with your finances. This is a freedom that you relinquish if you borrow from family or friends. A difficulty also can occur if you cannot repay the loan. You may lose a friend or family member. Consider these issues carefully before borrowing from family or friends.

Conventional loans are from a savings or commercial bank and are easy to obtain, provided that you have three items in your virtual possession: a good credit score, an income, and some equity in your residence. Your lender will need to confirm your ability to repay the loan. The three items just mentioned provide the lending institution with some assurance that you will repay the loan. The process is usually short, and you can obtain the funding quickly. The repayment terms usually are set and not negotiable. However, you can shop for loans from different institutions.

One type of financing that is often available is contractor financing. Your local contractor or home superstore may be willing to finance your home remodel. The question is, "Why?" Your service provider will benefit from this arrangement in three ways:

1. He or she will sell you a product.
2. He or she will sell you a service.
3. He or she will earn interest by financing your project.

Is this type of loan for you? The answer is "Yes" only if you understand all the terms of the loan and have them in writing. Many box stores or large remodel firms will finance your home remodel project for a limited time at a low introductory rate, but that rate may end soon after contract inception. These are individual businesses financing your projects. They must follow the lending laws, but they are not required to provide you with the best or even a good deal. These loans can be very competitive; they also can be very expensive. Only enter into a loan agreement when you have full comprehension of the terms, and you have those terms in writing. If you cannot pay for the home-improvement project without a loan, put a clause in the contract with the builder stating that the agreement is contingent on financing. This will allow you to cancel the contract if you cannot obtain financing.

A type of loan that you use every day is your credit card. I do not recommend that you use a credit card as a long-term financial instrument. Credit-card terms can be changed at any time for any reason or for no reason at all. You do not want to being paying 33.9 percent interest for your home improvements. If you are using your credit cards and intend to pay the bill entirely at the end of the billing cycle or when your loan is approved, this is an acceptable reason to temporarily finance your project with credit cards.

Low-interest loans typically are sponsored by the government for some type of specialty improvement, such as bailing out the U.S. banking system. In your situation, energy savings improvements to your home qualify for government assistance. For homeowners, the Federal Housing Administration (FHA) often sponsors such loans. Your state or local governments may have similar programs. Your local bank will be familiar with these loans and will be able to help you apply for funding. The "Green Revolution" has created opportunities for some home improvements that may be financed with FHA loans. The FHA does require an approved contractor, though.

Low-interest loans also are available for energy improvements in your home, new energy-efficient construction, and many other purposes. If you are planning a large home renovation, visit http://portal.hud.gov for the FHA and www.hud.gov/ for the U.S. Department of Housing and Urban Development. These Web sites explain your options in detail. These types of loans do come with some restrictions. These restrictions include the loan amount, limited scope of funding, and other issues. These types of loans are created to address and improve social issues. Energy efficiency has become one of the most significant social issues of our time.

The U.S. Department of Housing and Urban Development site is the portal to American Recovery and Reinvestment Act funding. Funding for many home-related improvements is available (Figure 10-4).

Zero-interest loans are also available as government-sponsored loans to help to address and resolve social issues. The rebuilding after natural

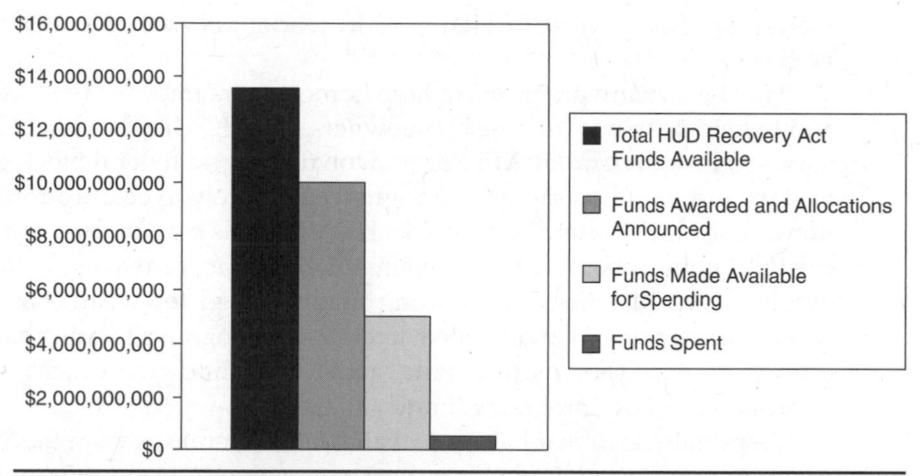

FIGURE 10-4 Public funding available. *(U.S. Department of Housing and Urban Development, May 2009.)*

disasters, the rebuilding after unnatural disasters such as September 11, 2001, low-income improvement programs, and currently, large-impact green improvement programs all qualify to be funded. The Internet will be your best resource for such programs. Federal and state governments will be the primary locations to find these types of loans. Your local government occasionally will fund such projects to better the community.

The funding available at the federal level comes from a few different agencies. HUD is a federal agency, but it will provide you with federal, state, and local information. The Web link www.hud.gov/local/ny/homeownership/homerepairs.cfm will link you directly to the State of New York and its programs for homeowners. Each state's program also is listed with HUD at www.hud.gov. HUD, however, provides more than just funding. At the HUD Web site you will find resources to help you better your home and your community. You can find financial counseling, predatory lending assistance, safety requirements for buildings, links to agencies for assistance, and much much more. The people at HUD are friendly and helpful.

A program presently available is the "Funds for Handyman-Specials and Fixer-Uppers" located at www.hud.gov/offices/hsg/sfh/203k/sfh203kc.cfm. This is called HUD's 203(k) program. This program is available to purchase or refinance properties. The program will allow a buyer to borrow the purchase price of a property plus the funding needed to rehabilitate that property. The loan is FHA insured and is provided through approved mortgage lenders in the area. You must have a good credit score to apply. The loan will include the purchase price of the home, the cost of repairs, and a 10 to 20 percent contingency reserve for unforeseen remodeling issues. A list of HUD 203(k) lenders is available at HUD's Web site.

HUD has many programs to help homeowners improve their homes and to help people to become homeowners. The HOME Investment Partnerships Program, or HOME Program, is the largest federal block grant to state and local governments designed exclusively to create affordable housing for low-income households. HOME funds may be used to assist existing homeowners with the repair, rehabilitation, or reconstruction of owner-occupied units. The funding may be used for weatherization, emergency repair, or handicapped accessibility programs to bring a property up to code. For more details, go to www.hud.gov/offices/cpd/affordablehousing/programs/home/.

A specialty loan that has been available to veterans is from the Veterans Association Program or, more commonly, the VA loan program. This is the loan program that is most famous for helping World War II veterans purchase their first homes. The U.S. Department of Veterans Affairs, often called the VA, is available at http://www.va.gov/. You can access the loan page directly at www.homeloans.va.gov/. The VA Web site includes all the information needed to direct you through the lending process.

Are you one of the few lucky people who cannot see your neighbor or his or her home? Then the next group of loans many be available to you. The U.S. Department of Agriculture (USDA) Rural Development Assistance Program is available to select borrowers. USDA's Rural Development Assistance is available for repairs of existing homes. The loans are available in rural communities and small incorporated towns of up to 10,000 people, but some communities of between 10,000 to 20,000 people also may be eligible. You can check the Web site at www.usda.gov to find out if you are eligible.

There are two additional types of federal loan options that deserve mention. *Deferred loans* are loans where the interest or principal or both can be deferred for a period of time. Most people are familiar with these types of loans because of their use primarily for education. These types of loans sometimes contain an option attached to government-sponsored loans. This option is called *loan relief*. These loans are "relieved" or "forgiven," that is, do not have to be repaid, when the conditions of the loan are achieved. Sometimes, when loans are provided to increase opportunities for homeowners or to improvement communities, if the "Expectation of Improvement" is met, then the loans are forgiven. These loans are unique, and all conditions must be met, but they are obtainable.

On a state and local level, you can find many types of services to help promote and benefit you and your community. A good example of homeowner assistance is from the Connecticut Department of Social Services

(DSS). The DDS administers the Connecticut Weatherization Assistance Partnership with utility companies and local community action agencies. The program assists low-income families with incomes up to 200 percent of the federal poverty guidelines to reduce their energy bills by making their homes more energy efficient. For more information about this program, go to www.ct.gov/dss/site/default.asp.

Home-equity loans are excellent options for people who own their homes. Using the equity in your home is quick and easy and often provides tax benefits. A home-equity loan is a loan that is in addition to your mortgage. The loan can be open or closed; that is, with an open loan, you can borrow and repay capital as you require. A closed loan is similar to a mortgage; you borrow a fix amount of funding and repay that funding on a prearranged schedule. Check with your lender and your accountant for tax considerations before you make a financial commitment.

A final loan option is to remortgage, or refinance your current mortgage. If you own your home and do not have a current mortgage, you will have no difficulty getting a loan. You can borrow as much as 80 percent of the current value of your home. If you currently have a mortgage, the option to borrow is the same, provided that you have equity in your home.

What is equity? If you have a home value of $500,000 and you currently owe $100,000 (your current mortgage), you can borrow 80 percent of the total value of the equity you have in the home. Thus total equity available is $400,000, so 80 percent of that is $320,000, which you would be able to borrow.

In some cases, you can borrow any amount up to the full amount of your equity in your home. So why can't you borrow 100 percent of the total value of your home? The mortgage lender/holder, usually a bank, requires 2 percent to as much as 10 percent to resell a home if the home goes into default, that is, you cannot pay your mortgage. In addition to these banking expenses, if the mortgaged home drops in value, then the owner will have borrowed more money than the home is worth. If the borrower of the loan defaults, the bank will lose money. Does this sound familiar? This was the primary cause of the worldwide financial crisis in 2009.

The benefits of a new mortgage are numerous. You can choose the terms that best suit your needs. You can borrow the amount that you require. A new mortgage will allow you to shop for a lender, and this gives you the ability to select the best rate and loan terms for you. You can "lock in" a fixed rate that cannot be changed. Finally your mortgage interest is tax deductible. Using the equity in your home is often the best solution for home improvements.

Once again, a great resource for all types of funding is the Internet. International funding in almost all countries is competitive and available on the Internet. Traditional banks, Internet banks, direct lenders, and other purveyors of secured loans can be found on the Internet. Internet sites such as LendingTree.com gather multiple quotes for you after only one brief, nonbinding application. These resources can provide you with viable vendors and competitive rates. You can compare rates and fees at your convenience in the comfort of your own home.

Nontraditional Sources of Funding

Funding your home does not always mean money advanced for your home remodel. You can use credit tools, provided that you have the ability to repay in the short term. Here are some examples of payback projects that can finance themselves:

1. Tax rebates
2. Tax credits (tell your accountant)
3. Grants you must apply
4. Self-funding projects such as solar, wind, or geothermal

Tax rebates are simple in design and allow the homeowner to take advantage of spending his or her money for better-quality products and services. To take advantage of tax rebates, identify the product or service that offers a tax rebate. Purchase the eligible product or service that you require. Then apply for the tax rebate. Applications for tax rebates are prepared and submitted separately or completed with your income tax return. Tax rebates are given to individuals regardless of tax status.

Tax credits are different from tax rebates. Tax credits are applied to offset tax balances. If you have no or a low income, you receive no benefit from a tax credit. Tax credits, like tax rebates, are designed to be simple. Eligible products and services will allow the purchaser to claim a tax credit. The same rules apply for tax credits as for tax rebates: Find the product or service that you require, be sure of its eligibility, and apply for the credit. Tax credits are applied at the time of your tax return. Tax credits and rebates affect your tax return, so be sure to mention these purchases to your tax accountant when applying for your refund.

Grants for energy products and services are available to average consumers. While the grant process may seem overwhelming, it actually is not. All grant information is delivered in detail. The requirements for receiving "free money" are strict and time-consuming but very much obtainable. I recommend that you begin by asking your local government if such funding exists at the local level. Local funding is easier to obtain, and you can work with the town or city to provide officials the proper information to achieve success. Local grant money is often provided to improve the efficiency of a home's doors, windows, roof, insulation, etc. These items are commonly paid for by such grants.

Your second option is to contact your state. Your state will have grant programs available on its Web site. This is where you begin your search for information. When you have found a program or grant that may fit your circumstances, follow the directions and apply for the grant. If you should need assistance, there is usually a grant administrator. This person is a state employee who overseas this particular grant. You can contact the grant administrator; he or she is always helpful. The administrator is there to help you understand the grant and application process. However, the administrator is not there to do the application for you. You must do your own work. This said, you can find a professional grant writer. These are people who most importantly understand how the grant process works and can explain to you the facts of each grant. Hiring a grant professional can be expensive. If you are pursuing a large amount of funding, you may want to hire such a professional to write your grant application.

A final public option is federal grants. All federal grants can be located through one Web site: www.grants.gov. This Web site contains tens of thousands of grants for every imaginable subject. Finding the grant that is perfect for your situation will require time and effort on your part. If you find a grant for which you are eligible, you should apply. For many of these grants, you can get assistance from the grant administrator or information officer. You will be on your own with many others, however. The research for and writing of a proper application will be time-consuming. You must understand that the government is thorough because it is literally giving away money. The process cannot and should not be made easy. That's called *communism*. If you find that the time required for the application is prohibitive, then you should look for similar programs at the state and local levels.

Your final option is private grants. Yes, there are many types of organizations that give away funding: not-for-profit organizations, foundations, and trusts. All these groups were set up to give money away. That said, much as with the federal government, the application and then the selection process can be very difficult. The Internet will be your best research tool to find many of these organizations. Who would give away funding? On such group is the Carnegie Foundation. Andrew Carnegie was once the richest man in the world. He gave away most of his money to establish many libraries, schools, and universities in America. Founding the Carnegie Corporation of New York, Carnegie Endowment for International Peace, Carnegie Mellon University in Pittsburgh, and the

Carnegie Museum of Pittsburgh, Andrew Carnegie, like most wealthy individuals, created organizations that would continue to promote subjects important to the endower.

There are countless foundations and other organizations that fund specialty interests. More modern-day philanthropists are funding green technology projects. Check your local list of billionaires to see what they are doing. For example, Al Gore has joined Kleiner Perkins Caufield & Byers (KPCB), a venture-capital firm located in California. The key to finding private funding is to find a person or organization that has the same interests as you.

The final way to fund your home remodel projects is with *capitalism*. What do I mean? Let's begin with solar power. You can place a solar power system on your roof that will provide you with more electricity than you can use. What happens to the extra electricity? You sell that electricity back to the utility company or to other private buyers. You don't believe me? Read my book on solar power (available summer 2010). Let's say, for example, that you produce $10,000 worth of electricity but you only use $5,000 worth. You now have $5,000 worth of "extra" power. Since you produced the power, you sell the power, and you receive payment. This type of project can produce not only the funding to pay for itself but also an income for you once you have paid for the project.

Depend on your geographic location, you can perform these types of projects with solar, wind, geothermal, and many of the other forms of renewable energy. Land rental, placing multiple synergistic green energy projects together, etc., are all possibilities. Speak to a renewable energy expert or your local utility. These organizations will be able to help you to determine if these projects are the best solution for you. You also can find quality resources at www.exploresynergy.org.

You may now have completed this book, but I would bet that I have led you to many new questions and hopefully many solutions. If you choose only to do the no- and low-cost projects in this book, you should be able to reduce your energy consumption and therefore increase your income by 30 percent. This is not bad. When was the last time you received a 30 percent raise at your place of employment?

Saving energy, creating your own power, and just living smarter is the way of the future. If you do not care about the future, then why not live more comfortably and have more income to spend now?

Index

I

K

L